Anatomy
in Action

Anatomy in Action

The Dynamic
Muscular Systems
that Create and Sustain
the Moving Body

BY **THEODORE DIMON, EdD**

With illustrations by **G. David Brown**

Foreword by **David I. Anderson, PhD**

North Atlantic Books
Berkeley, California

Published by
North Atlantic Books
Berkeley, California

Cover art and illustrations by G. David Brown
Cover design by G. David Brown and Jasmine Hromjak
Book design by Happenstance Type-O-Rama

Printed in the United States of America

Anatomy in Action: The Dynamic Muscular Systems that Create and Sustain the Moving Body is sponsored and published by North Atlantic Books, an educational nonprofit based in Berkeley, California, that collaborates with partners to develop cross-cultural perspectives; nurture holistic views of art, science, the humanities, and healing; and seed personal and global transformation by publishing work on the relationship of body, spirit, and nature.

North Atlantic Books' publications are distributed to the US trade and internationally by Penguin Random House Publishers Services. For further information, visit our website at www.northatlanticbooks.com.

MEDICAL DISCLAIMER: The following information is intended for general information purposes only. Individuals should always see their health care provider before administering any suggestions made in this book. Any application of the material set forth in the following pages is at the reader's discretion and is their sole responsibility.

Library of Congress Cataloging-in-Publication Data
Names: Dimon, Theodore, Jr. author.
Title: Anatomy in action : the dynamic muscular systems that create and
    sustain the moving body / by Theodore Dimon.
Description: Berkeley, California : North Atlantic Books, [2021] | Includes
    bibliographical references and index. | Summary: "A guidebook
    illustrating the principles behind the body's musculoskeletal system"—
    Provided by publisher.
Identifiers: LCCN 2020036965 (print) | LCCN 2020036966 (ebook) | ISBN
    9781623175801 (trade paperback) | ISBN 9781623175818 (ebook)
Subjects: LCSH: Musculoskeletal system—Anatomy. | Human locomotion.
Classification: LCC QM100 .D558 2021 (print) | LCC QM100 (ebook) | DDC
    612.7/6—dc23
LC record available at https://lccn.loc.gov/2020036965
LC ebook record available at https://lccn.loc.gov/2020036966

1 2 3 4 5 6 7 8 9 VERSA 26 25 24 23 22 21

TABLE OF CONTENTS

FOREWORD

have admired Ted Dimon's work since I read one of his earliest books, *The Elements of Skill: A Conscious Approach to Learning*, many years ago. My admiration and enthusiasm for his work only increased after I read this latest book. The book is an important contribution to the literature and it has the potential to positively impact the field of human movement science as much as the profession of Alexander Technique teaching. Moreover, it offers a practical self-help guide to the countless numbers of people who seek a means for dealing with the myriad of maladies that result from chronic stress and tension.

I write this foreword several months after the start of the COVID-19 pandemic. The human species has experienced multiple pandemics over millennia, but one is particularly relevant to the subject matter of this book. It is the pandemic of musculoskeletal ailments that besets modern humans; a pandemic that may have started when our ancestors transitioned from hunter gatherers to farmers some twelve thousand years ago, but which was certainly exacerbated and accelerated by the start of the industrial revolution just over 250 years ago. F. M. Alexander argued convincingly that modern civilization was responsible for much of the physical deterioration observable in modern humans. Through systematic self-observation and experimentation, he developed a technique to address this deterioration. As proponents of the technique know, it is a reeducational technique rather than a therapeutic technique, treatment, or cure. It rests on the premise that specific problems in the musculoskeletal system can be understood and addressed only in the context of how the system operates as a functional whole. It stresses the importance of the dynamic relations among the parts of the system, not the least of which is the dynamic relation between the head, neck, and back in determining the integrated functioning of the whole system.

Alexander taught us that a generalized expansion of the musculoskeletal system characterizes effective and efficient movements. Such expansion is particularly apparent in a lengthening and widening of the torso and in the forward and upward movement of the head relative to the spine. He did not, however, have an anatomical or physiological explanation for why lengthening and widening was so fundamental to movements, though he was masterful in bringing about such lengthening and widening in his pupils. This is where Ted Dimon makes his primary contribution. He advances a theory of how the musculoskeletal system works that not only clarifies many of the teaching procedures in the Alexander Technique, but also challenges scientists to think about human movement in a new light. The theory builds on the principles of systematic self-observation and experimentation advocated by Alexander and draws on decades of experience teaching the Alexander Technique as well as recent scientific advances in our understanding of human biology.

Contemporary anatomy books provide detailed descriptions of bones, articulations, ligaments, and muscles in a mechanistic and reductionist manner. Yet they fail to explain how the entire collection of components that make up the musculoskeletal system has evolved to function as a coherent whole. These books privilege understanding the parts over understanding the whole, perhaps because we have such a poor understanding of what binds the parts into the whole. *Anatomy in Action* takes a very different approach. Ted Dimon contends that the musculoskeletal system can be understood only in the context of its function to provide upright support. That support is not simply supplied by the bones, as is often taught to students of anatomy, but instead is a function of the dynamical relations between all of the components that comprise the musculoskeletal system, including bones, muscles, and the entire network of connective tissue. Counterintuitively, Ted Dimon argues that a critical contribution to the integrity and stability of the system, as well as to heightened proprioceptive awareness, is the muscles' capacity to lengthen between their bony attachments. When the system functions the way it evolved to function, movements are light and effortless. When we interfere with that natural functioning, problems arise. These ideas are sure to cause a buzz in the scientific community and cause many to rethink the merits of the traditional and contemporary *corrective exercise* approaches to improving musculoskeletal functioning and human performance.

A lucid description of how the parts of the musculoskeletal system relate to the whole and how the whole in turn affects the parts is not the only contribution *Anatomy in Action* makes to our understanding of human movement. Equally important are the clear and concise step-by-step procedures for restoring the integration between the parts and the whole. The rationale behind the procedures, many of which will be familiar to teachers of the Alexander Technique, becomes obvious when viewed in the context of the theory of musculoskeletal functioning advanced in this book. Moreover, the procedures reveal it is possible to treat the root causes underlying musculoskeletal system dysfunction rather than just the symptoms associated with such dysfunction.

This book is beautifully illustrated and replete with photos, diagrams, and examples that highlight and clarify its contents. Ted Dimon's latest work is a valuable resource for anyone who seeks to relieve muscular tension, restore lost function, improve coordination, or achieve peak performance for themselves or others. It is also a wonderful read for anyone who is curious about how modern humans can effortlessly control an amazingly complex body by thinking clearly about its natural design. I will consult this book frequently in my own work!

David I. Anderson, PhD

Director, Marian Wright Edelman Institute for the Study of
Children, Youth and Families, San Francisco State University and
Professor, Department of Kinesiology, San Francisco State University

INTRODUCTION:
The Key to Our Human Anatomical Design

Few people today, it seems, are free from joint, back, or muscle pain. With our increasingly sedentary way of life, these problems are bound to become worse. Children, generally speaking, are free from musculoskeletal problems, but huge numbers of adults in today's stressful world live with musculoskeletal difficulties of one kind or another.

To a large extent, these problems can be attributed to our modern lifestyle, coupled with the distinctly human upright posture that has made this lifestyle possible. The spine of a four-footed animal is oriented horizontally and supported by fore and hind limbs. To produce upright human posture, the trunk must be brought directly over the hind limbs, which gives us a generally unstable form of support that is prone to various forms of collapse. This upright posture also requires the development of a lumbar curve, which makes the lower back highly unstable and vulnerable.

Given this scenario, it seems we must do one thing if we want to overcome tension problems: balance faulty postural habits by removing tensions and compensating for the weaknesses associated with these imbalances. Today, dozens of methods are designed to do just that—virtually all of them based on the premise that our posture is faulty and must be corrected.

As I will show in this book, however, this approach tells only part of the story. Our upright posture is amazingly subtle and complex. The notion that "if something goes wrong, we must correct it" completely overlooks the capacity of the body to function naturally and healthfully. Why, for instance, do most young children, and even some adults, exemplify perfect coordination and muscle tone, without exercise or intervention of any kind? The answer is that, when working properly, the body is designed to function in a balanced and effortless way and, furthermore,

is designed to function this way at a completely automatic level when not interfered with. To properly treat the problem, we need to understand how the muscular
system is designed to work, because therein lies the secret to how we can function
normally and healthfully.

Instead of understanding this design, however, we rush to correct the problem
by massaging, stretching, releasing, or treating muscles. If, for instance, we experience pain while sitting, the usual assumption is that muscles are working incorrectly
and that we must correct these imbalances by strengthening muscles or balancing
the activity of opposing muscle groups. This approach, however, treats symptoms
and only superficially addresses the real problem. If, instead, we take time to examine and restore our natural ability to sit, we will not only be able to sit easily and
effortlessly, but in the process, we will reinstate natural and automatic changes in
muscular and reflex functioning that are far deeper and more meaningful than what
we can achieve through corrective exercise, treatment, or massage.

"Physician, Heal Thyself!"

When we suffer from musculoskeletal difficulties, we usually assume that something is wrong with the body and, in response, seek the help of professionals—
massage therapists, physical therapists, and so on—who possess objective knowledge about the body and how it works. The ability to cure disease based on objective
study of the body is the foundation of the biological and medical sciences. If I am
suffering from an infectious disease, I go to a doctor who has the knowledge and
experience to identify and treat the disease based on the study of this disease and
the forms it takes. In somewhat similar fashion, various experts are trained in the
diagnosis of posture, muscle function, and exercise physiology based on the clinical
study of patients and ways of treating them.

As I will show in this book, however, muscle tension cannot be diagnosed or
treated in the same way that medical scientists study disease. If, for instance, you are
suffering from lower back tension, a physical therapist or massage therapist may be
able to identify and treat tight muscles and, in this way, provide therapeutic relief.
But the solution to tight muscles—which, after all, does not constitute a medical
condition in the normal sense of the word—is to restore the lengthened working
of muscles based on an understanding of the coordinated working of the back and
related muscles. No amount of treatment can produce this condition, which requires
knowledge of how the back works, how to bring about length in muscles, and how
to coordinate the body in action. As I will show in this book, this kind of knowledge
requires a new paradigm and new ways of looking at the body and how it works.

This is not to say that clinical diagnosis and medical treatment are never warranted in musculoskeletal problems. If you woke up this morning with shooting pains in your shoulder and arm, it would be not only reasonable but advisable to find someone with the medical and clinical expertise to identify the specific problem and, if necessary, to treat it. Symptoms such as these suggest that something is wrong, and when something is wrong, you need to go to the appropriate expert. But muscle tension as a general problem belongs to a different category of function and requires a new kind of knowledge about the coordinated working of the musculoskeletal system and the ability to coordinate muscles based on this knowledge. This kind of knowledge cannot be gained through objective or clinical observation alone, any more than the anatomical study or dissection of the larynx can teach us to command the proper use of our voice.

Beyond Healing, Treatment, and Cure

Belief in treatment and cure is deeply embedded in our ways of thinking about and approaching the human body. If something is wrong—or if we *think* something is wrong—we want someone to tell us what the problem is so that it can go away. In such cases, we rely on the authority of experts who have specialized knowledge of the disease in question. Such an attitude may be warranted when we are ill and require treatment, but it is not nearly as appropriate in addressing muscular tension. In this case, we must understand how the body works in a positive way and how to bring about these improved conditions on our own through a process of education and self-study. Forms of treatment not only fail to provide this knowledge but, even worse, reduce the subject to being a passive recipient of treatment. This is true even of holistic techniques, many of which, in the name of education, treat symptoms without imparting real knowledge to the subject about how the body works and how to perform actions in a more balanced and conscious way. To be valid, a system of kinesthetic education must be based on principles that can be learned by a motivated student, that can be applied in a meaningful way to daily activity, and that lead to ongoing insight and knowledge about the body and about oneself. This kind of know-how requires a new kind of science based not on forms of treatment but on observation and knowledge of the living organism in action.

History

My own exploration of this subject began many years ago, when I injured my back in college. While stretching after a workout, I felt a twinge and, soon afterward, my back went into spasm and I was immobilized for several hours. I thought I had simply pulled a muscle but, within a few weeks, the spasms were happening every few days. I assumed that, because I had done a lot of strenuous exercise, my muscles were too tight, so I tried a number of methods for reducing tension—relaxation, stretching, yoga, meditation—but to no avail. Even when my muscles were relaxed, I noticed that, once I resumed normal activity, the problem returned. I began to suspect that I was doing something harmful and began to explore kinesthetic methods that would heighten my awareness of what I was doing in action.

One such method had been developed by F. Matthias Alexander, an Australian actor who discovered that his own vocal and breathing difficulties were caused by a harmful pattern of tension that interfered with how his body was designed to function naturally.[1] Based on his observations, he developed a practical method for helping people notice and prevent these tensions. I began having lessons in the Alexander Technique and found that my back problem was connected with harmful tensions in my back and legs and that, by paying attention to these tensions, I could get some relief. Wanting to learn more, I began to train full time in the Alexander Technique and, within months, began to experience periods where the tension was entirely absent but would return when I succumbed to my old habits. It soon dawned on me that nothing had been wrong with my back and that my problem had been caused by unconscious habits that interfered with the natural functioning of my muscular system. Interested now in the subject as a whole, I went to graduate school, where I began to explore the problem of performance, habit, and action as a new subject in educational development. I also wanted to introduce the subject of awareness in action as a new field of study, to start an institute, and to train teachers in the new field.

To do this, however, I needed to clarify various aspects of the subject on which these views were based. First and foremost was the notion that the body had a natural design—a claim for which I had little evidence or support. If, for instance, you look at anatomy books, they can tell you a lot about particular muscles, but how the parts relate to one another—or whether an overarching design principle explains how the parts organize as a whole—is completely omitted, as if the body works in parts and the whole is simply the sum of these parts. Yet clearly there was a connection between body parts—one that provided the key to how muscles and the musculoskeletal system work as a whole. When I began training teachers, I observed the coordinated working of this system every day in my students. Yet clear as this knowledge was to me at an experiential level, I could not account for it in a concrete and coherent way.

The breakthrough came when I realized that muscles function on the basis of length. The usual views are that muscles contract to produce movement and support and that if muscles are tight or shortened, they can be stretched, massaged, or released but cannot actually lengthen. What these approaches missed is that, to support the skeletal framework, muscles are actually lengthened between their bony attachments—a concept that is completely counterintuitive because muscles that are lengthening (and not eccentrically contracting) cannot support anything. And yet this is what I actually observed in people: muscles produce support not by shortening but by lengthening—a quality in muscles that seems to function not simply as a remedy for tight muscles but as a fundamental principle in nature. Based on this, I developed a theoretical model for how the musculoskeletal system works—or what I called the *postural neuromuscular reflex (PNR) system*—that included muscle length, muscle tone, and the organizing relationship of the head and trunk. The description of these elements, and how they function as a whole, became the foundation for my book *Neurodynamics*,[2] in which I lay out a basic theory and practice of the control of the organism in action.

Having established this theoretical model, however, I realized that I still had not described in detail how the musculoskeletal system worked. I knew, for instance, that the shoulders are normally shortened and narrowed and, for the system to work properly, they need to widen. This involved releasing the flexor muscles that cause the narrowing, which requires a process of paying attention to the muscles and giving them time to release. But I now realize that it is possible to go much further than this. As a system, the shoulders have a definite design, and each of the components of this system—the flexors in front, the extensors in back, the rotator cuff muscles—participates in this larger system in a clearly defined way. The same is true of the musculoskeletal system as a whole, which is made up of a number of subsystems that can be described in detail.

The description of these systems, which I began as an exploratory project but which has since become increasingly concrete and defined, is the basis of this book. Various methods offer to cure or treat various problems in the body without addressing the much more fundamental question of how the body is actually designed to work. As I worked on this project, I began to lay down a detailed anatomical description of each system and how it works. This begins, as we will see in Chapter 1, with a review of the basic theory of musculoskeletal function (the PNR system) and, in Chapter 2, with an explanation of how this theory applies to the different systems in the body. In subsequent chapters I look at the specific anatomical systems—extensors and flexors, the shoulder girdle, the upper limb, and so on—and describe in detail how each works. In this way, I not only show that we possess a naturally coordinated musculoskeletal system, but I describe how each of these systems contributes to the working of a coordinated whole. Having now worked with thousands of people over nearly four decades, in practice I have seen how each of these systems can be restored, based on a working knowledge of their dynamic design and how to restore it. Neither the identification of postural faults nor forms of exercise and treatment are substitutes for practical knowledge of how these structures work as a coordinated, dynamic system.

It is my hope that my present work, in describing this system in detail, will contribute to a practical and theoretical understanding of this vital subject and to an advancement in our knowledge of the remarkable and subtle workings of that greatest of all instruments, the human body.

The Musculoskeletal System and Its Dynamic Design

Although it is an obvious fact that muscles produce movement, how the musculoskeletal system works as a coordinated whole is far less clear. Although we are all familiar with the idea of exercising and stretching muscles to keep them healthy, the notion that the body works as a complex, dynamic structure designed to function naturally is less familiar to us.

The basic premise of this book is that the musculoskeletal system—muscles, bones, and connective tissue—forms a complex structure that works dynamically to support us against gravity, and understanding how this structure works is the key to healthy musculoskeletal function.

In this chapter—and in the pages that follow—I present a new and revolutionary model of how the body works based on practical and concrete knowledge of muscles, bones, and connective tissue and how they cooperate to form a complex, coordinated whole.

The Body in Action

As a moving machine, the human body is one of evolution's greatest marvels. Our ability to walk on two feet, to run, dance, climb, and throw, to make things with our hands, and to speak makes us the most skilled movers on the planet. Although many of these skills require learning, we acquire them—and are motivated to learn them—at a largely instinctive level. By the age of five or six, the typical child is capable of an enormous variety of skilled actions, refining and expanding on these skills for decades to come.

When we are young, we learn actions with relative ease, and the musculoskeletal system—even when we have not fully mastered the use of it—seems to function effortlessly. Not so as we get older. By the time we are in our thirties, most of us cannot sit in a chair, or use our arms while sitting at a computer or playing an instrument, for more than a few minutes without discomfort. Increasing numbers of young people complain of back pain, and musculoskeletal conditions have become the leading cause of disability worldwide. Although we become more skilled with age, the quality of our actions tends not to improve but to deteriorate.

But what accounts for the difference between the easy, poised grace of young children and the awkward, debilitated actions of many adults? Because we experience specific problems, we typically attribute the cause to the immediate problem—ruptured disc, strained muscles, injured shoulder—and treat it. Aren't strain and discomfort an unavoidable consequence of work, stress, and life in general? Young children spend all their time at play; as teenagers, we are forced to sit in classrooms for hours on end. Life is stressful; as we age, we lose muscle mass, our tissues deteriorate, and we should not expect our bodies to function as well as they did in childhood.

As we'll see in this book, however, the reason children enjoy such natural functioning is not simply because they are young and healthy but because their musculoskeletal systems function naturally as a whole. We have only to look at how a child stands and sits to see that their system works as a coordinated whole. The body as a whole is supported naturally and easily, all the muscles are lithe and toned, and the various parts work together to produce effortless support against gravity. In the normal adult, these elements are missing because the elements that make up the whole are working improperly based on harmful patterns of action that, over time, interfere with this system.

The notion that there is a natural movement system and that this system is the basis for healthful, efficient functioning is something that many of us intuitively understand. Yet how—or why—the body works in this way has never been fully articulated or understood. The different parts of our anatomy—shoulders, back, hips, and so on—make up a dynamic relationship of parts that, when organized properly, are designed to function easily and effortlessly. Making sense of how this system works as a coordinated whole is the subject of this book.

The Dynamic Role of Muscle

It is tempting, when we speak about the musculoskeletal system, to look first and foremost at muscles. To perform a movement—to lift a glass, to walk down the street, or to type a letter—we have to contract, or tighten, particular muscles; otherwise we would not be able to move in space, manipulate objects, speak, or even breathe. The ability to move is organized by the nervous system, which sends signals to muscles that cause them to contract and other signals that tell muscles not to contract. At least, this is the usual view of what muscles do. What complicates the situation is that, in order to move in space or even to move an arm, we first have to keep ourselves upright in the field of gravity—in other words, we have to maintain postural support. Most of us have heard about the postural muscles that keep us upright—the deeper muscles of the neck, back and spine, and legs. By contracting, we are told, these muscles keep the head from toppling forward, the trunk from buckling, and the legs from collapsing under us.

But exactly how do postural muscles work? When you lift a heavy book, muscles in your shoulder and arm forcibly contract, moving the levers of the arm and maintaining the weight of the book. Using a great deal of force to accomplish the task is a perfectly acceptable strategy because you don't have to hold the book up for very long. If your arm tires, you can soon put the book down and rest your muscles. Maintaining postural support of the body is a different matter. To support the entire body in the gravitational field, muscles have to maintain the support of the trunk for hours at a time, and trying to do this by simply contracting muscles would lead to exhaustion and dysfunction. How, then, do muscles and bones actually work together to support the body as a whole?

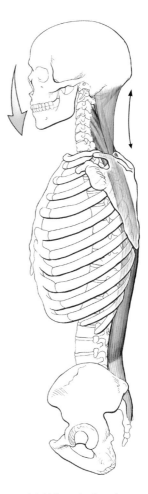

The answer is that they don't simply contract but work in a dynamic partnership with bone. The muscles on the nape of the neck, for instance, must maintain the support of the head, which would otherwise fall forward. But they don't perform this function simply by contracting and pulling on the head, which would cause the head to be constantly pulled back and fixed in place. Instead, the head is weighted in front so that it exerts an opposing force on the neck muscles and keeps these muscles lengthened. The neck muscles can then pull on the head but, instead of forcibly pulling it back, they are stretched between

Fig. 1-1. When the head is allowed to balance naturally, the muscles of the spine can lengthen.

the head and the spine so that, even while the muscles contract, the skeleton maintains length in the muscles (Fig. 1-1).

Another structure that exhibits this dual tendency is the spine. Muscles must pull on the spine but, if this was all they did, the spine could not maintain its lengthened support against gravity but would be dragged down by the muscles. Instead, muscles are lengthened between vertebral attachments and ribs and, in this context, maintain the support of the spine (Fig. 1-2). This arrangement explains how muscles act on bones as a basic way of supporting the skeleton. In order to produce force, muscles must contract—something we see every day in humans and other vertebrates. To maintain basic postural support, however, muscles must lengthen between the bony structures to which they attach so that, even as they act on bones, they are lengthened between bones—a subtle and dynamic arrangement in which muscles and bones work together to produce support against gravity. Some variation of this relationship exists in virtually every part of the musculoskeletal system. Instead of simply contracting, the muscles are suspended within a latticework of bones in such a way that, even while they maintain the stability of the skeleton, they are lengthened between the bony structures they support.

Fig. 1-2. Muscles don't simply act on the spine but are lengthened between attachments, forming a complex rigging for upright support.

The Body's Tensegrity Design

If muscles don't simply pull on bones but are lengthened *between* bones, how then can we explain posture and musculoskeletal support? According to the traditional biomechanical view, bones form a supporting framework upon which muscles act to maintain posture and to produce movement. In this model, muscles contract to support (and move) the bones, and length is not a key part of the equation. There are two cases in which muscles do in fact lengthen: when muscles are eccentrically contracting (as when you lower yourself by your arms), and when they are passively lengthened (as when you stretch your calf muscles). In the first case, the muscle is still actively contracting; in the second case, the muscle is being acted upon and is not performing a useful function. Muscles still perform only one function, which is to contract, or shorten.

Here, however, we see something quite different. Muscles do not appear to actively contract at all but instead seem to lengthen between bony contacts, and yet they support body parts. How is it possible that muscles can lengthen rather than contract, and yet produce upright support? The answer is provided by the concept of *tensegrity*. In a tensegrity structure, no fixed supports hold things up, just guy wires and struts, yet these two elements, working together, are all that is needed to produce a self-supporting structure. In this model, muscles don't act on bones but are embedded within a latticework of muscles and connective tissues. According to this quite different kind of model, upright support is achieved not by the bones, but by the larger network within which bones float.

A simple example of a tensegrity structure is the old-fashioned tent (Fig. 1-3). This tent has no supporting walls, just the central pole. The pole holds up the walls of the tent but, at the same time, the walls of the tent support the pole. In this arrangement, opposing forces create support—the pole resisting the pull of the tent walls and the tent walls acting as tensile structures that resist stretch and keep the

Fig. 1-3. A tent with a single pole and guy wires is supported through tensegrity principles.

pole from falling. In more complex tensegrity structures, the poles, or struts, float within a network of guy wires, creating a supporting structure held up mainly by the guy wires and not the struts (Fig. 1-4). In this case, the struts float within the network of guy wires, providing no direct support, yet miraculously the structure seems to maintain support against gravity. In the human design, bones float within a network of connective tissue and muscles carry most of the load, creating a flexible, supportive structure that is both efficient and mobile.

Tensegrity structures are often compared to more traditional structures such as columns, arches and walls, which are designed to resist compression and to bear weight. Whether made of bricks, girders, blocks of stone, dried dirt, or concrete, compression structures have been used for centuries in the construction of cathedrals, coliseums, temples, aqueducts, and houses of all kinds. A tensegrity structure, in contrast, combines rigid compression members and tensile members to produce a strong, lightweight structure. The word *tensegrity*—a combination of *tension* and *integrity*—is a term coined by Buckminster Fuller, who developed the concept.[1] In tensegrity structures, the rigid members don't bear weight but provide opposition to the tension members, which in turn pull on the compression members. A *tensegrity*

Fig. 1-4. Geodesic domes, such as the biosphere in Montreal, are tensegrity structures.

structure, then, can be defined as a continuous tensile network, interspersed with struts that create framing against which the tensile elements pull. In such a structure, much of the work is borne by the tensile members, which distribute the strain evenly throughout. This makes for a very efficient design that is lighter and stronger than walls or beams and uses less material.

Because tensegrity structures are artificially engineered utilizing tensile wires and struts, they appear to be advanced and hi-tech. In comparison to those found in nature, however, such man-made structures are in fact rather crude. The tensegrity dome shown here utilizes regularly spaced struts and guy wires; in animals we find vastly more sophisticated structures, such as the rabbit pictured here (Fig. 1-5). In humans, this design reaches a pinnacle of complexity and subtlety that makes it possible for us to perform an amazingly diverse array of activities while distributing the workload over many meters of connective and muscle tissue; this way the burden does not fall on just a few muscles but is borne by all. Specific muscles perform work, but in the context of a larger, dynamic system in which bones are suspended between tensile structures, enabling the system as a whole to maintain support in the gravitational field with a minimum of effort and strain.

Since Buckminster Fuller popularized the concept, tensegrity has been widely studied and, in the newly applied field of biotensegrity, is being used to model living

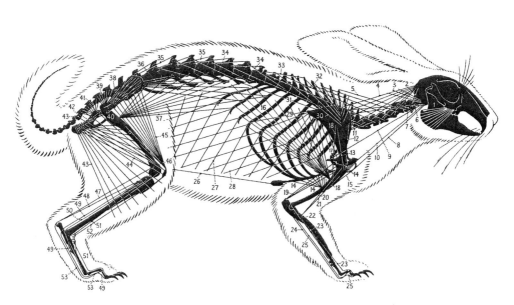

Fig. 1-5. A line drawing of the musculoskeletal system of a rabbit reveals complex tensegrity elements.

organisms.[2] At the simplest level, it is now clear that muscles do not act as singular structures but are, in fact, part of a larger network of fascia and connective tissue. One key role of fascia is to distribute directional forces and to transmit these forces across bony boundaries.[3] Good examples of this, which we'll look at in detail in Chapter 6, are the spiral muscles that wrap around the trunk. Because the oblique muscles of the trunk are separated in front and back at the midline, forces are transmitted across the midline by means of fascial networks to form larger systems that wrap continuously around the body (Fig. 1-6).

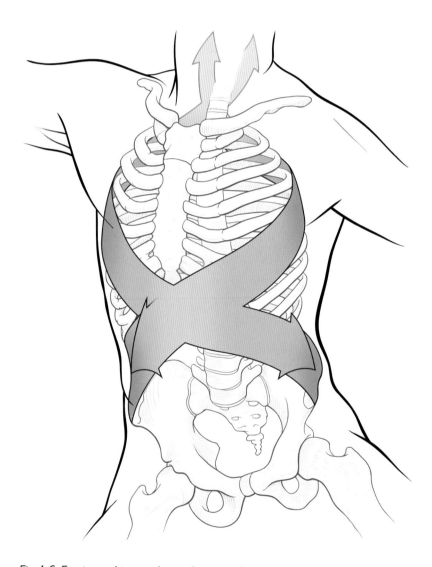

Fig. 1-6. Fascia combines with muscles to produce continuous lines of force, shown here in the trunk region.

Until fairly recently, fascia was thought to be a rather inert structure, but it turns out that fascial tissue possesses its own tensegrity architecture, sometimes referred to as the *extracellular matrix*.[4] This viscoelastic structure, like the larger tensegrity architecture of the body, resists deformation and, at the same time, distributes and responds to deformation over the entirety of its structure. At an even deeper level, the cells within the extracellular matrix are prestressed tensegrity structures that respond holistically to pressure.[5] Thus the tensegrity design of the body at the macro level (connective tissue, muscle, and bone) is supported by tensegrity at the micro levels, constituting a nested, multifractal tensegrity design.

The Contractile Function of Muscle

In our account of musculoskeletal function, muscles contract to produce movement, but they also work in a coordinated fashion to produce total bodily support in a gravitational field. Within this larger framework, muscles do not simply pull on bone but lengthen between their bony attachments and, in this context, maintain tone in order to create postural support. Therefore, to form a complete picture of how muscles actually work, we must describe two elements of muscle function: how they contract to perform work, and how they lengthen between bony contacts as part of their supporting function. The contractile function of muscle, which we'll look at first, is well understood and, as we'll see, is essential to understanding the second function.

Muscle fibers are long, thread-like cells bundled together to form muscles, which are attached to bones to produce movement. In highly specialized organs, like the eye, muscle fibers are a fraction of an inch long, and only a very few are sufficient for the job at hand; in other parts of the body, such as the thigh, these fibers can be up to two feet long, and many thousands are bundled together to form powerful, bulky muscles that make it possible to walk, jump, and run on two feet (Fig. 1-7).

Just as a muscle is made up of many individual muscle cells, or fibers, each muscle cell is made up of tiny, thread-like *myofibrils*, which are the contractile units within each muscle cell that make it possible for the muscle to perform work. Within each fibril are two types of molecular chains—the thin actin and the thick myosin chains—that are stacked together in such a way that the two types of filaments interdigitate, giving the muscle the banded or striated pattern after which striated muscle is named (Fig. 1-8).

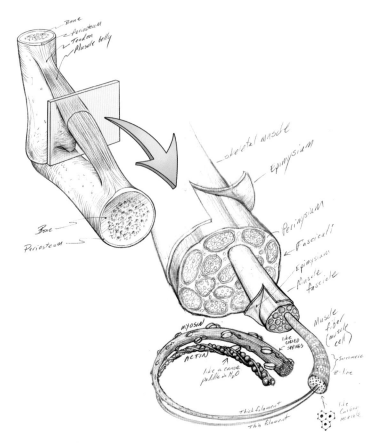

Fig. 1-7. The cross section of a typical muscle reveals bundles of individual muscle fibers, which themselves are composed of myofibrils.

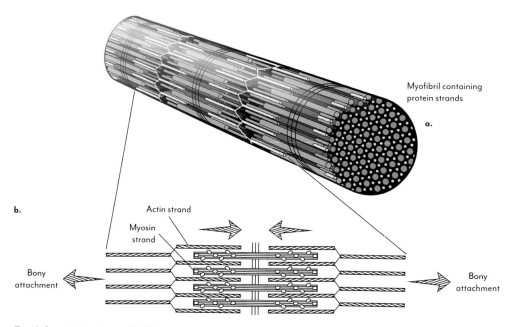

Fig. 1-8. a. A single myofibril showing an ordered patterning of myosin and actin strands; b. The interdigitation of the myosin and actin strands accounts for the muscle's striated appearance.

It is the ratcheting movement of the thin filaments in relation to the thick ones that produces contraction of the muscle, which is set off by chemical activity within the myofibrils. The myosin filaments have regularly protruding globular heads that are capable of binding at particular regions on the actin filaments. In a relaxed muscle, the myosin heads do not come into contact with these sites on the actin molecules because they are blocked by troponin molecules. In this state, the myofilaments are able to slide alongside each other, allowing the muscle to be passively lengthened. When the muscle fiber receives a signal from its motor nerve, however, this releases calcium ions into the myofibril space, which bind with the troponin molecules, changing their shape and exposing the binding sites along the actin molecules. The globular heads of the myosin molecules now bond at regular intervals along the actin molecules, forming cross-bridges between the actin and myosin strands (Fig. 1-9).

One of two things can now happen. If the myosin heads remain bonded along the actin strands, the cross-bridges create stiffness or tone in the muscle because the interdigitating strands are now linked together and cannot slide alongside each other. The muscle then resists being lengthened—exactly what we see when a muscle is not entirely relaxed but maintains stiffness or tone, which maintains postural support.

The second possibility, of course, is that the muscle can actively contract or shorten. In this case, the myosin heads only momentarily form cross-bridges because they continue to bond at adjacent sites along the actin chain, rotating and drawing the actin chain along the myosin chain (Fig. 1-10). This happens because, when the myosin head attaches to the actin filament, it pivots, or rotates, which moves the actin filament in relation to the myosin filament. After each rotation, the myosin head detaches, straightens itself out, reattaches at a new point on the actin chain and, through this continuous ratcheting action, draws the actin chain along the myosin chain. This telescoping of the actin and myosin strands, which can

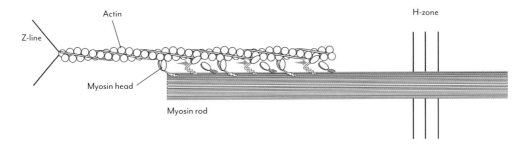

Fig. 1-9. The globular heads on the myosin chains bond with the actin chains to form cross-bridges.

be seen on an electron microscope as a narrowing of the striations in the muscle, causes the muscle fiber as a whole to shorten, or contract. All of this is what we would describe as normal muscle function. Muscle contraction is produced at a molecular level when the motor nerve sets into motion the chemical changes that cause the interdigitating strands of myosin and actin molecules to slide over one another and thus shorten the muscle.

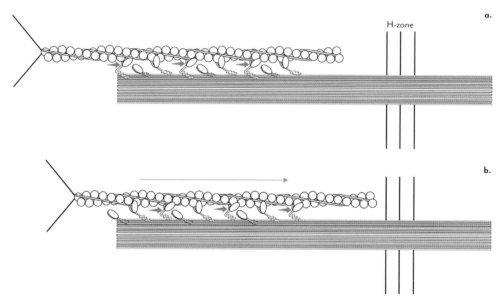

Fig. 1-10. a. Chemical changes in the muscle allow the myosin heads to bond with the actin chain; b. The ratcheting action of the myosin heads along the actin filament causes it to move along the myosin chain toward the H zone.

The Second Function

So much for the contractile component of muscle function; let's look now at how muscles lengthen between their bony attachments. We've seen that, when a muscle contracts, the interdigitating strands of molecules within the myofibrils slide alongside each other to contract or shorten the muscle, shown here in schematic form (Fig. 1-11). In contrast, muscles that perform basic postural duties (such as those at the back of the neck) do not actively shorten but lengthen between their bony contacts. This, of course, is not something they "do" but, as we've seen, it happens naturally as part of how muscles work with bones to form a supportive structure.

But exactly what does it mean for muscles to lengthen, and how does this differ from muscle contraction? Consider what happens when we slump in a chair and

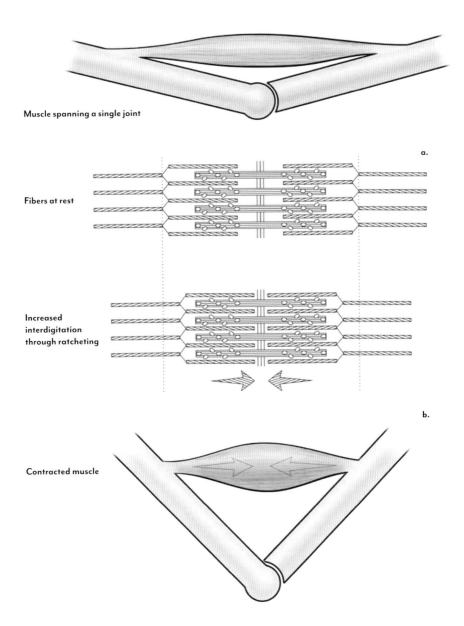

Muscle spanning a single joint

a.

Fibers at rest

Increased
interdigitation
through ratcheting

b.

Contracted muscle

Fig. 1-11. a. At rest, the fibers in the muscle cell are lengthened; b. As the muscle fibers shorten through ratcheting action, the muscle contracts and acts on the bone to produce movement.

the head is pulled back by shortened neck muscles. In this posture, the extensor muscles of the back are inactive and, because the spine lacks the support it needs to maintain its length, the trunk collapses into the slumped state. With the head pulled back in this way, natural upright support is virtually impossible because, instead of exerting an upward force on the neck muscles, the head is pressing down on the spine and the spine is unable to lengthen.

To restore postural support, we must stop pulling the head back so that it is not actively pressing down upon the spine and the trunk can come to its natural length. But restoring length is not simply a matter of making postural adjustments or stretching shortened muscles. In order to lengthen, the neck muscles must first stop contracting, which in turn allows the head to move upward, shown here again in a schematic way that corresponds to how a four-footed animal's head is balanced at the end of its spine (Fig. 1-12). When this happens, the muscles are able to assume their natural length in

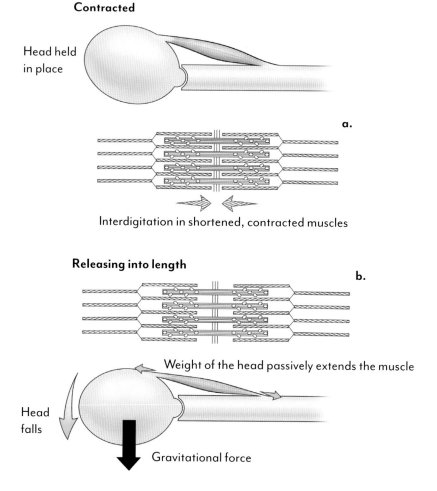

Fig. 1-12. a. A heavy body part (such as the head) can be held by shortened or contracted muscles; b. When a muscle stops contracting, gravity intervenes and the part drops; the muscle is stretched and the actin and myosin strands slide apart.

the context of their bony attachments, triggering stretch reflexes that cause them to tone up and to resist being lengthened further (Fig. 1-13).

To address the practical problem of muscle tension, then, we have to understand how muscles that are contracting must release or let go so that they can lengthen between their bony attachments. For this to happen, the cross-bridging that takes place during contraction must cease so that the molecular strands are able to slide apart from each other. Understanding this dual function of muscles is an essential

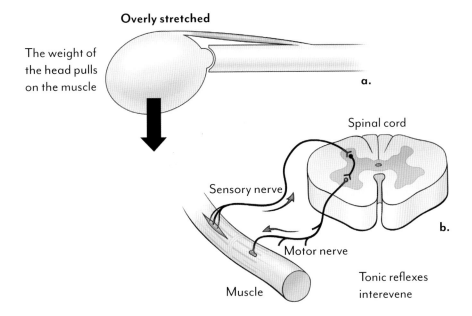

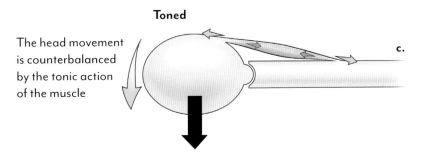

Fig. 1-13. a. When gravity acts on the head, the head in turn pulls on the muscle, which becomes over-stretched; b. Stretch receptors in the muscle trigger tonic reflexes that counter the gravitational force; c. When the tonic reflexes are triggered, motor signals to the muscle enable it to resist further stretching.

part of what it means for muscles to be healthy. Muscles that are chronically contracted must be able to stop contracting so that they can actively lengthen between their bony attachments. This is a dynamic state that exists only in muscles that are actively performing their functions within a dynamically supported, bony framework. When working in this way, muscles take on a healthy, spongy quality that can be felt to the touch—neither overworked, as in chronically shortened muscles, nor flaccid, as in muscles that have become weakened because they are no longer serving their tonic function. Producing this condition of muscle tissue is not simply a matter of releasing muscles by treating or stretching them; muscles lengthen in the context of a skeletal framework and, in this context, are designed to maintain natural length, as we see in this image of a young child (Fig. 1-14).

Fig. 1-14. The natural poise of young children as they move their upper limbs to play is indicative of healthy lengthened muscles and tonic support.

Proprioception and Muscle Length

To summarize what we've said so far, the body is not a simple bony framework acted upon by muscles but comprises a complex tensegrity system in which bones, connective tissue, and muscles cooperate to form a supporting structure. What has stood out in this model is the fact that, instead of pulling on bones in a crude manner, muscles actually lengthen between their bony attachments, creating an interactive structure that distributes forces throughout the system, and in such a way that the entire system works as a coordinated whole.

For this system to work, however, it is not enough for muscles to be lengthened; in order to support the skeleton, they must also maintain tone or stiffness in response to the stretch exerted by the bony members of the skeleton. How do they do this? Part of the required force is produced by the nature of the tensegrity system itself. In the traditional view, muscles are motors that pull on bones, but in a tensegrity system, muscles pull inwardly against the outward pull of the struts, making the system pretressed.[6] When the connective tissues are lengthened even further by adding greater load, the tensile structures store kinetic energy, which further increases the stiffness and resistance in the muscles—something that can be felt, in a healthy and naturally prestressed system, as liveliness and bounce.[7] This ability of muscle tissue to firm up and gain efficiency in response to stretch appears to be built into the design of muscle tissue itself. Within their myofibrils, muscles contain strands of proteins known as *titin*; when lengthened, these molecular strands respond by firming up the muscle, adding force with no added metabolic demand being placed on the muscle.[8] In short, the tensile component of muscle and connective tissue, acting within the context of the body's tensegrity architecture, contributes significantly to the tone and strength of the system.

Muscle tone is also maintained by higher centers in the brain. When I lift my arm, the movement is produced by the contraction of muscles that are activated by motor nerves. A nerve impulse, which begins at the higher cortical centers of the brain, carries a motor message to the muscles, which is combined with other nervous impulses to produce coordinated movement. But this isolated activity takes place in the context of a larger system of muscular support. If, for instance, I raise my arm, many muscles are involved in the support of the shoulder girdle and in the postural support of the body as a whole—a process that is far too complex to be directed piece by piece. We never just contract one muscle; the entire support system must constantly adjust itself in relation to whatever we are doing as the background against which specific contractions take place.

This overall support, which produces what we all know as *posture*, is the work of stretch reflexes. In this context, we can no longer speak of muscles as discrete organs or as parts of a larger musculoskeletal system; muscles are part of a complex proprioceptive system designed to sense, and respond to, changes in this system. When our muscles become tight and sore, it is easy to become preoccupied with how tight they are and to assume, accordingly, that the solution is to release them or treat them. Virtually all myofascial systems—yoga, bodywork, massage therapy, and so on—address muscle tension in this way, as if muscles are unruly children that require our constant supervision. We forget that muscles are controlled by the nervous system in the most subtle and remarkable ways and that, furthermore, they are invested with sensory organs that constantly send feedback to the nervous system, comprising a system of remarkable subtlety and sophistication.

This suggests that there is a great deal more to the problem of muscle tension than the fact that muscles become too tight and require stretching. And yet, for the most part, we have studied these functions—the motor nerves that tell muscles to contract and the muscle spindles that give information about muscles—as a sort of fixed system that simply does what it needs to do. If something goes wrong with our muscles, we assume they are simply shortened and require the aid of a massage therapist or yoga instructor; we have little to say about what has gone wrong (or even if it has gone wrong) and how the system can function more efficiently. When we understand the dynamic role of muscle tissue within the bony framework of the skeleton, however, we can begin to say something much more definite and meaningful about muscles and how to address muscle tension. When we restore length to muscles, they respond by toning up, at an entirely automatic level, in response to this length. If, for instance, shortened back muscles begin to lengthen, they now do less work and support the trunk more efficiently, paradoxically producing greater effect with less effort. Muscles all over the body function in this way as part of a larger system that maintains support with minimal work.

But why should muscles that are letting go provide support? To put it another way, why should muscles contract more effectively when they are letting go? Shouldn't letting go mean that, instead of providing more support, they provide less? The answer is that muscles are designed to respond to length—that is, they are wired in such a way that, when they lengthen, the muscle spindles, which are designed to sense stretch, send automatic signals to tell the muscle to contract at a low-level, constant rate—otherwise known as muscle tone (Fig. 1-15). What this means, in practical terms, is that muscles don't need to lengthen simply because they become too tight and need to relax. They are meant to lengthen as part of how they perform their supporting function and, when they do, they tone up accordingly. Releasing a muscle by stroking or massaging may provide relief, but it cannot achieve real lengthening and in fact tends to override this system by desensitizing and over-stretching the muscle

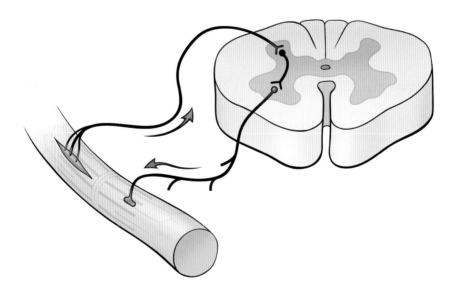

Fig. 1-15. A tonic reflex is generated when a spindle in the muscle responds to stretch by sending a signal to a motor nerve serving the same muscle.

tissue. It is only when muscles actively lengthen as part of a dynamic system that the stretch reflexes are activated and the muscles tone up in response.

Length is also the basis for heightened proprioceptive awareness. Trying to kinesthetically sense what is going on in our muscles as the basis for making direct changes may seem constructive, especially when we associate kinesthetic awareness with forms of release that relieve strain and tension. But awareness has little value if it cannot be applied in activity, and the only way to become aware in action is to establish natural muscle length based on the coordinated working of the system. When the body works as a dynamic whole, we awaken the kinesthetic sense in new ways, and can begin to sense how we interfere by shortening muscles. In this way, we can establish a meaningful foundation for becoming aware of what we are doing in activity. Establishing length in muscles thus serves as a crucial foundation for becoming proprioceptively aware in the context of a holistic dynamic system over which we can gain greater control.

The sensitivity of stretch receptors to dynamic length in muscles tells us that the nervous system is not a static structure that regulates the musculoskeletal system but is, in turn, influenced by the working of this system. The nervous and musculoskeletal systems interact in a reciprocal relationship in which the nervous system regulates muscle tension, and proprioceptive function is in turn directly dependent on how the musculoskeletal system functions as a dynamic, working whole. Understanding this makes it possible to translate what we know about neural function into meaningful terms that, as we will see in this book, can be applied in a very practical way to muscles and sensory awareness.

The Central Role of the Head and Spine

We have seen that, for the body to work as a total system, muscles must work in a dynamic partnership with bone to produce lengthened support. This is not only an efficient way for muscles to maintain the postural support of the skeleton, it is also how muscles register changes in length so that motor nerves, in turn, can maintain proper muscle tone. Muscles, working in conjunction with the skeletal framework, thus function as a central component of a neural and motor system of remarkable subtlety and complexity.

But not all muscles are the same. When we lose postural support and go into a slump the head is pulled back and exerts downward pressure on the spine. To restore natural support, the neck muscles must lengthen so that the head can balance forward, removing the downward pressure and allowing the spine to lengthen. It is not a coincidence that we are focusing here on the head. Sitting atop the spine, the head is the highest point in the body, which means that, to maintain upright posture, the head must not be pulled down but move (for want of a better word) in an upward direction. In humans, the head leads the body as a whole, and the spine, in turn, has to lengthen to maintain the upward support of the body as a whole. In other words, the head and trunk form an essential foundation for movement.

This central role of the head and trunk applies even to the use of the limbs. If, for instance, you reach with your arm to pick something up, you must engage specific muscles of the shoulder and upper limb. But the muscles of the shoulder and arm never work in isolation from the head and trunk, which must be supported and stabilized as the basis for using the arms. When we use the limbs, the muscles supporting the head and trunk form the essential foundation for the use of the limbs and are, in this sense, central to all other movement.

Even when we are simply sitting and using our arms, the head/trunk relationship—and our tensegrity design as a whole—is fundamentally connected with movement in space. As vertebrates, we evolved from marine creatures that possessed a tubular gut with a mouth at the front end (Fig. 1-16). Muscles along the length of the body moved the fish in a forward direction, based on sensory input received at the front end (Fig. 1-17).

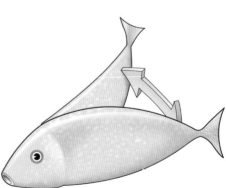

Fig. 1-16. The precursors of vertebrates were marine animals that propelled themselves through the water by laterally flexing the body.

Fig. 1-17. Fish possess a rigid spine which enables more efficient movement in a forward direction.

This arrangement changed quite dramatically in fish that used their fins to move about in the shallows and then began to crawl onto land. In this case, the fins became functional levers for moving on the ground. To move on land required new kinds of support against gravity. In a marine environment, a fish is buoyed up by the water and therefore lives in a world that is virtually gravity-free. Animals on land, however, have to contend with the pull of gravity; to move efficiently in space, they must first raise themselves off the ground. Some reptiles accomplish this task with their bellies close to the ground and their legs splayed out to the sides; mammals walk and run with their legs more directly underneath them, making it possible to move quickly and efficiently on land. Here, the spine comes into play in a completely new way, functioning as a kind of bridge for supporting the body on the forelimbs at one end and the hind limbs at the other (Fig. 1-18). The musculoskeletal system of terrestrial animals thus serves a dual function: first, it produces overt movement by acting upon bones as levers; second, it counteracts the tendency of these levers to buckle so that, as the animal moves about, the body can be supported on its four limbs. The overt function of muscles in producing movement now takes places against the background of a prestressed, toned tensegrity structure maintained by muscle tone, organized around the head and trunk relationship, and moving directionally in space.

In humans, this horizontal design has been modified for upright balance on two legs. This places the spine, as well as the muscles that support it, in a vertical arrangement. The system now lengthens upward against gravity and not in the direction of movement; but all action—even when it just involves sitting and using the arms—is still organized around the head and trunk, which lengthen upward as part of our upright design (Fig. 1-19).

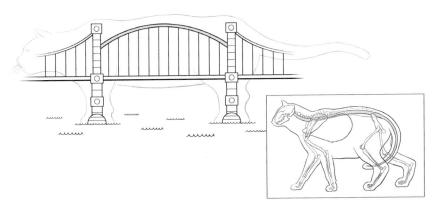

Fig. 1-18. In mammals such as a cat (inset) the spine forms a bridge between the fore and hind limbs.

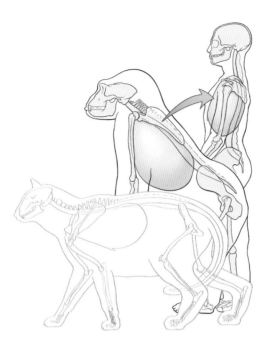

Fig. 1-19. In humans, the horizontal spine of a four-footed animal has evolved to a vertical position.

DYNAMIC OPPOSITION AND THE PRINCIPLE OF EFFORTLESSNESS IN NATURE

The notion that muscles pull on bones to produce movement is a basic biomechanical concept going as far back as Leonardo da Vinci, whose drawings of bones and muscles as a kind of lever/pulley system (Fig. 1-20) provide an enlightened account of the natural mechanics of movement. Yet such a view cannot explain how action takes place in a living, breathing animal for two reasons. First, when muscles contract, this action pulls bones together, which creates fixation and loss of mobility of the parts in question. Although specific muscles can act on particular parts in this way without detriment to the whole, if all the muscles did this, the entire structure would lack flexibility and fluidity. In order for vertebrates to move efficiently, the entire skeletal framework must be fluid and mobile, and contractions would disturb this overall fluidity. Second, muscle contraction requires a great deal of energy, and a living organism with limited resources cannot handle such a high metabolic demand. To meet these two requirements, moving structures are built upon tensegrity principles that lower energy expenditure by allowing muscles to remain lengthened while, at the same time, enabling relevant muscles to contract without compromising the fluidity of the whole.

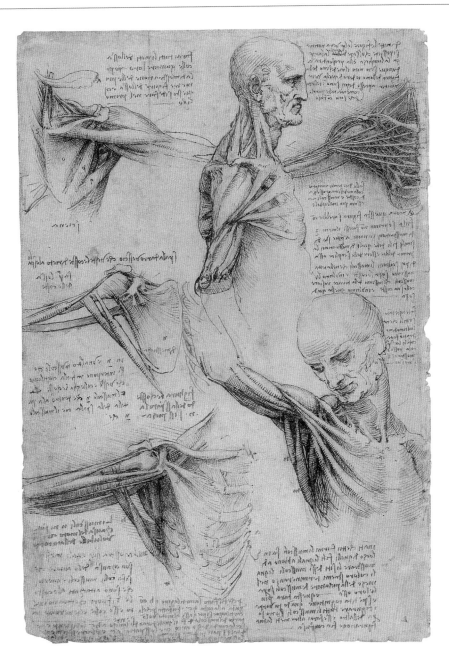

Fig. 1-20. A Leonardo da Vinci drawing depicting the biomechanics of muscle and bone.

In order to work in this dynamic way, major segments of the body do not move toward each other but apart. When we forcibly contract a muscle, one part of the body is pulled toward the other; in the context of the body's larger tensegrity architecture, body parts move in opposition. This creates a dynamic polarity that operates as a basic principle in vertebrate design, as we can see in the cat pictured in Fig. 1-21. Particular muscles are forcibly contracting, yet the neck muscles, instead of shortening, are lengthening in such a way that the head goes away from the spine and the spine lengthens as a whole. The same thing happens in humans except in a vertical orientation: the extensor muscles of the neck and back are active but, instead of shortening the spine and pulling the head toward the spine, the neck muscles lengthen so that the head goes up and away from the spine and the spine lengthens (Fig. 1-22). In other words, muscles don't simply pull on bones; connectives tissues and muscles interact dynamically with bones to produce a supporting structure that is fluid and mobile. This is a basic principle in vertebrate movement in which a polarity of forces creates directional support in space with a minimum of fixation and effort. It is nature's solution to the problem of vigorous and efficient movement, a kind of yin/yang of bodily motion and support.

Fig. 1-21. In this frame-by-frame image of a leaping cat, we can see that the entire body lengthens as the cat prepares to jump.

Fig. 1-22. In effortless, efficient sitting, the head balances naturally on the spine and the muscles of the neck and back lengthen.

Once we understand how this system works, we can see that forms of movement or treatment that address muscular tension in purely structural terms cannot establish the dynamic conditions required for efficient animal movement. Structure is important, but no amount of structural work can establish these dynamic relations, which depend upon the dynamic working of the system as an energized, living whole. The same can be said of exercise, which works on the misguided assumption that the main function of muscles is simply to pull on bones and completely leaves out the dynamic quality of muscle function. Both approaches are unidirectional, mechanical concepts, and nature does not work on the unidirectional or mechanical principle; it works on the principle of dynamic opposition and polarity of forces. Compression presses outward and muscles pull inward. The combined action creates direction of parts with a minimum of effort—a principle we see everywhere in nature, exhibited as litheness and effortlessness in action (Fig. 1-23).

Fig. 1-23. The effortless stretching and twisting of this agile cat is indicative of litheness in movement as a principle in nature.

The same principle applies to the concept of relaxing muscles to reduce tension. Because so many of us suffer from excess muscular tension, it is easy to think that release of tension is the main quality we're looking for in muscles. But muscles do not need simply to let go or release; they must let go in the context of a skeletal framework in which parts move in opposition, producing the dynamic complementarity required for mobile yet forceful movement. Achieving this quality is not just about muscle release but about directional energy. When, for instance, the neck and back muscles stop tightening in the context of dynamic length, the head moves quite forcefully in an upward direction, which it needs to do if it is to counteract the force of gravity and the falling weight of the body. When this happens, the head actually pulls in an upward direction, producing energetic support and directional movement based on the two-way polarity of tension and stretch. In cats, as we just saw, this lengthening takes place in the same direction as its movement in space; in humans, the oppositional lengthening force is upward because, although we move forward in space when we walk, we are oriented vertically and lengthen upward against gravity.

THE SPINE AND ITS TENSEGRITY DESIGN

A fascinating example of the concept of tensegrity support was reported some years ago in the *New York Times*. It compared European men carrying heavy loads on their backs to Kenyan women carrying weight on their heads (Fig. 1-24).[9] The women, it turns out, carried 20 percent of their body weight with no additional expenditure of calories as compared to the men, who used far more effort. The study concluded that this was because the women altered their gait but did not alter their upright support mechanism when carrying a load on their heads, whereas the European men did and therefore had to use far more muscular effort to support the packs on their backs. In essence, each woman was able to carry the weight on top of her head by utilizing—and not disturbing—the tensegrity support of the musculoskeletal system, so that the load was distributed over the entire tensegrity structure rather than straining particular muscles.

Fig. 1-24. Women carrying weight on their heads with a minimum of work.

It has been long accepted that to support weight, the spine, which acts as a compression structure, must be acted upon by muscles that were viewed as the main agents in upright posture. It is now clear that as a supporting column, the spine is incapable of supporting heavy loads and that muscles can account for only a fraction of the force required to support heavy loads. What has been left out of the traditional equation is the role of fascia, which transmits and distributes forces throughout the skeletal framework.[10] This new model suggests that the spine is not simply a compression structure but a complex tensegrity system in which fascia not only transmits forces but actively contributes to the support and stabilization of the spine (Fig. 1-25 and Fig. 1-26). This tensegrity model explains why it is possible to carry heavy loads on our heads without damaging the vertebrae and without straining muscles: bones are embedded within a fascial network that not only transmits and distributes forces but also relieves muscles of work.

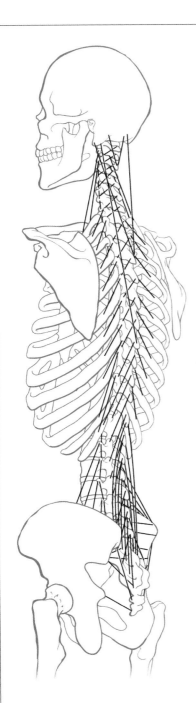

Fig. 1-25. When key muscles of the spine are depicted as lines, a complex tensegrity structure emerges.

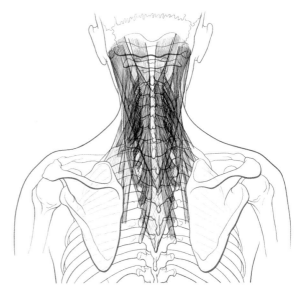

Fig. 1-26. The tensegrity patterning grows in complexity as more muscles are added, forming an elaborate system of rigging.

How the System Works

2

I t is a basic premise of this book that the musculoskeletal system is designed to work naturally and effortlessly, if only we take the time to understand this design and how it works. When something goes wrong with our muscles, we typically identify the specific symptom—a tight or strained muscle, inflammation, repetitive strain injury—and treat it with massage, strengthening, or stretching exercises. But the system works as a coordinated whole, and when we misuse a part, we misuse it as part of this whole. Treatment can be useful, but if we want to restore the natural and healthful function of the musculoskeletal system, we must understand how this system is designed to work and how muscles function in the context of this larger system.

The idea that, when something is wrong, we should work on muscle and connective tissue that has become in some way distorted reflects the more general attitude that the body is supposed to work properly; if it doesn't, we presume that something specific has gone wrong and it's the job of an expert—doctor, chiropractor, physical therapist—to correct it, like a car that breaks down and gets brought to the mechanic. But the body is a complex, finely tuned instrument and, if you are experiencing constant backache or muscle strain, it means that you have been doing something to interfere with this system over time. To sit comfortably, for instance, the muscles of the back must be able to support the trunk, the spine must be naturally lengthened, and the hips must be free and released. When we habitually slump, all these systems become interfered with, and no amount of stretching or strengthening will correct these imbalances if we don't understand how these systems are designed to work as a dynamic whole.

As we will see in the chapters that follow, ten key systems contribute to the working of the musculoskeletal system; we will look at each of these systems, examining how the system in question is designed to work, what goes wrong with it, and how it functions properly. In addition, we will look at some basic ways we can encourage the proper working of each system.

With this in mind, and before we look at the individual systems, it is useful to list a few of the key principles that govern the working of these systems.

1. Each system has a functional design.

When we suffer from particular musculoskeletal difficulties, we are usually so pre-occupied with addressing specific muscles or joints that we are unlikely to consider how the parts in question are designed to work. When, for instance, you are sitting for long periods and your back hurts, how are you using your back? Are you trying to sit up and, if so, are you overworking your lower back? The truth is that, if your back hurts, there is probably a good reason, and the solution is not to stretch or strengthen muscles but to understand how the back is designed to function in its supportive role. Every system has a key organizing principle; in each case, we identify the muscles that make up the system and the essential principle that governs its proper working.

HOW THE SYSTEMS WORK

Here are some of the systems we will look at in this book in Chapters 3, 4, 8, and 9.

The Extensors of the Back

The back muscles can be flaccid or shortened, but it does not follow that by strengthening or stretching particular muscles, the problem will be corrected. The proper function of the back depends on the length of the back muscles in relation to head balance (Fig. 2-1).

Fig. 2-1. The lengthening support of the back.

The Flexor Muscles

The flexor musculature is a tensile structure designed to support the ribs and abdomen and is suspended between points above and below (Fig. 2-2 and Fig. 2-3).

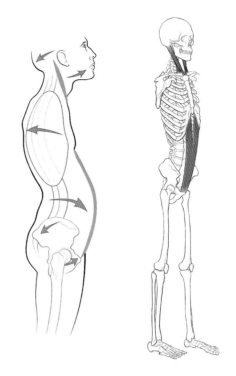

Fig. 2-2. Loss of front length.

Fig. 2-3. Restored front length.

The Shoulder Girdle

The shoulders can be raised, narrowed, collapsed, and fixed; to work properly, the muscles of the shoulder must be toned and lengthened so that the shoulders widen apart (Fig. 2-4).

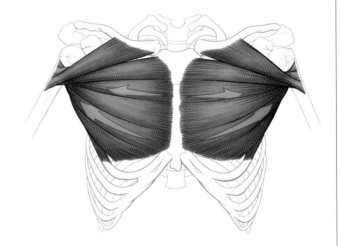

Fig. 2-4. The widened shoulder girdle.

The Throat and Its Suspensory Musculature

It is clear when there is a harmful degree of tension in the throat or when the tongue is harmfully depressed, as we can see in the photo (Fig. 2-5). But this does not tell us in a positive way how the throat is meant to function. To answer this question, we must understand how the throat is suspended from the skull and is thus maintained within a network of muscles that antagonistically support it (Fig. 2-6).

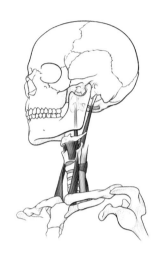

Fig. 2-5. Depressed larynx.

Fig. 2-6. Suspensory muscles of the throat.

2. Length is a key principle of muscle function.

Muscles are designed to contract, but they are also part of a larger tensegrity structure designed to produce support and movement. In this context, muscles are meant not only to contract but also to lengthen between their skeletal attachments, which is part of how the body maintains natural and effortless support (Fig. 2-7). The capacity of muscles to lengthen between their bony attachments is thus a fundamental property of healthy muscle tissue and the foundation for efficient support and movement.

Fig. 2-7. Natural sitting posture with muscles lengthened instead of shortened.

3. Muscle groups function in the context of upright support.

Although muscles in general are designed to contract to produce movement, specific muscles never function in isolation but are always part of a larger functional group, as in the case of arm movements, which involve torquing and spiraling the body. Even when we come to understand that individual muscles are part of functional systems, however, it is important to recognize that functional systems are themselves part of an even larger functional system designed to support us against gravity. Nowhere is this more true than with the spiral muscles, which are designed not simply to torque or twist the body but to maintain lengthened support against gravity, as we can readily see when we come out of a postural twist and the body releases into length (Fig. 2-8). What this means, in practical terms, is that muscles cannot be understood—or treated—individually or in functional groups but only in the context of upright support, which is the true basis for restoring normal function in muscles.

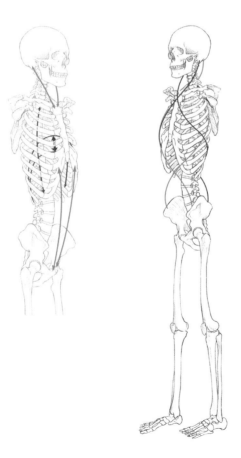

Fig. 2-8. The spiral muscles are part of our antigravity support system.

FUNCTIONAL MUSCLE GROUPS ARE PART OF OUR DYNAMIC UPRIGHT DESIGN

The spiral muscles of the trunk are designed to produce torsional movements of the body and, in this sense, have a clear functional purpose (Fig. 2-9). But this is not their only function. We have only to look at how these muscles attach to the skull to see that they are part of our upright design. For these muscles to function properly, they must lengthen in the context of the upright support system, as we will see in Chapters 5 and 6.

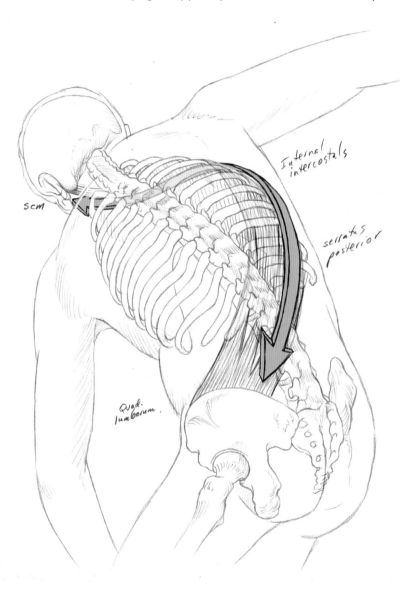

Fig. 2-9. Spiral muscles make rotational movements possible.

4. The relationship of the head to the trunk is the key organizing principle.

Although muscles contract to produce specific movements, such as flexing an arm or turning the head, movement in general is organized around the central relationship of the head and trunk. In four-footed animals, the head and spine are oriented horizontally to produce forward movement in space (Fig. 2-10); in humans, we are oriented vertically to maintain support on two feet, and all movement—whether forward locomotion in space or the fine motor use of the hands while sitting—is based on this upright support system, organized around the dynamic relationship of the head to the spine (Fig. 2-11).

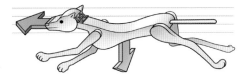

Fig. 2-10. Dog with head leading and body following in the direction of movement.

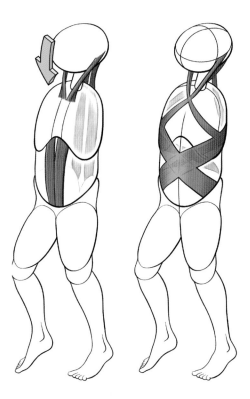

Fig. 2-11. Human with head and trunk in vertical arrangement.

5. The key to the part is the whole.

When we suffer from a specific problem, such as a painful shoulder, we normally identify the pain itself as being the problem or, if we attempt to make a more in-depth diagnosis, the inflammation or muscle strain that is causing the pain. This type of explanation may suffice in cases in which a specific problem requires treatment. When it comes to tension-related problems, however, this type of explanation tells us almost nothing, since it amounts simply to a description of the symptom, not an explanation of its cause. For instance, the working of the shoulder is entirely dependent on the working of the trunk, and shoulder problems are almost always part of a larger pattern of interference in the trunk. To address the shoulder, then, it is necessary not just to treat the specific symptom, but to restore the working of the larger system upon which the shoulder depends. In short, the whole is the key to the part, and a real understanding of how the different parts of the body work must be based on a practical understanding of this larger system.

6. The parts influence the whole.

As a system of levers, the hand and arm are dependent on the whole but not essential to it—or at least, that is how the arm, as a biomechanical system, seems to work. But each system in the body—arms included—is not only dependent on the whole but can also influence it in profound ways. If, for instance, the shoulders are narrowed and the arms are heavy and collapsed, the torso is adversely affected, not to mention breathing, muscle tone, and overall upright support. To address the body as a whole, it is necessary to restore the working of the shoulder and arm, which, in turn, restores the working of the upright system. This is why we cannot look at the musculoskeletal system as a whole unless we understand how the parts relate to the whole and how the whole is affected by the parts.

7. Many systems, one response.

Although we will examine individual systems in depth, it is important to remember that each one is part of a larger whole and contributes to the working of the whole. When working properly, these systems combine to produce effortless, expansive support of the body against gravity. Muscles lengthen between their bony contacts, the body is lightly and easily supported against gravity, and movements take place effortlessly and without strain in any part. This expansive movement, which involves the coordinated working of muscles all over the body, is the background against which all movement happens—the basic organizing principle of movement and the key to the musculoskeletal system (Fig. 2-12).

Fig. 2-12. Child with a naturally functioning upright support system.

How the Chapters Are Organized

Each chapter begins with a general description of the specific system being examined, how it is organized to work properly, and what goes wrong with the system, or what I somewhat playfully call *contraindications*. This overview is followed by black-and-white line drawings depicting the key muscles within the system, accompanied by a list of the origins and insertions of each muscle. We then examine the system in more detail, accompanied by numerous sidebars that explore specific aspects of the system. In examining each of these systems, the goal is not to list muscles and joints in exhaustive detail but to describe how the muscles work as functional systems. For a more detailed examination of musculoskeletal anatomy, see my earlier book, *Anatomy of the Moving Body*.[1]

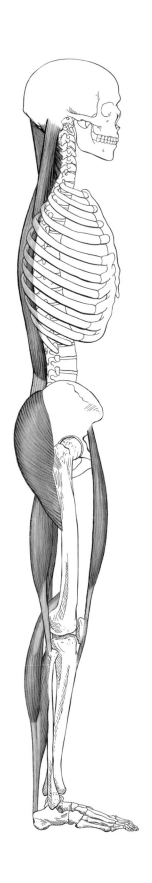
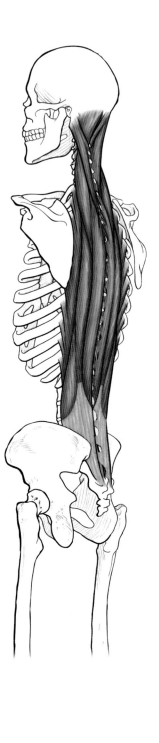

Extensors and Head Balance

3

The primary muscles that support us against gravity are the extensors of the back and legs. If you are standing and allow your body to go limp, you'll notice that as you begin to fall, your head, trunk, and knees buckle forward. In order to prevent this buckling, the extensor muscles, which lie primarily along the back of the body, counteract this tendency by extending the legs and trunk. Because extension at the knee is performed by the quadriceps muscles on the front of the thigh, the extensors of the leg are on the front of the leg as well as in back.

The Lengthening Back

When the muscles of the back are lengthened, the back maintains the support of the trunk with minimal effort. Acting as a counterbalance that opposes the pull of the extensors of the neck, the forward balance of the head on the spine is an essential component in maintaining stretch on this extensor sheet, enabling the muscles to provide support against gravity with a minimum of effort. In this context, the extensor sheet lengthens from the sacrum right up to the occiput, acting as a whole to maintain the support of the head and trunk.

Contraindications

Because we constantly pull the head back and shorten in stature when performing simple actions, the muscles of the neck and back become shortened. When coupled with the tendency to slump while sitting, which further disengages the extensor system, the back becomes harmfully narrowed and overworked in some places and flaccid in others.

WHEN WE SHORTEN IN STATURE

1. The neck muscles become shortened and the head is pulled back.
2. The back is narrowed.
3. The lower back muscles are shortened and contracted.
4. The chest is raised.

WHEN WE LENGTHEN IN STATURE

1. The neck muscles release and lengthen.
2. The head balances forward and up on the atlas.
3. The back lengthens and widens.
4. The back muscles become toned.
5. The trunk lengthens.

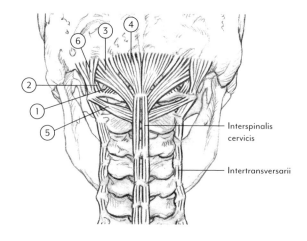

Interspinalis cervicis

Intertransversarii

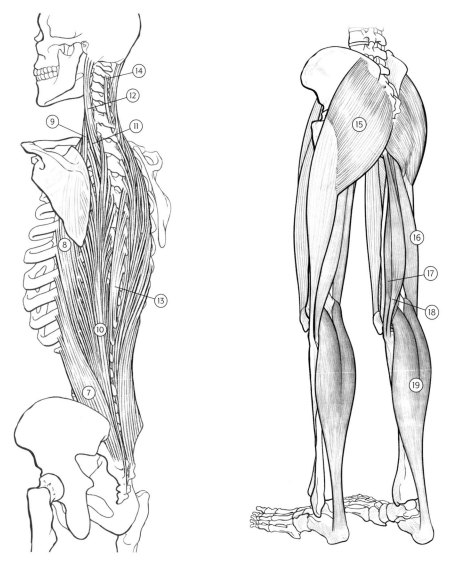

For vertebra and rib numbers listed in the table on the opposite page, see diagrams on pages 88 and 141.

ORIGIN	MUSCLE	INSERTION
Transverse process of atlas	**1. Rectus capitis anterior**	Occiput
Transverse process of atlas	**2. Rectus capitis lateralis**	Occiput
Spinous process of axis	**3. Rectus capitis posterior major**	Occiput
Atlas	**4. Rectus capitis posterior minor**	Occiput
Spinous process of axis	**5. Obliquus capitis inferior**	Atlas, transverse process
Transverse process of atlas	**6. Obliquus capitis superior**	Occiput
Sacrum/iliac crest	**7. Iliocostalis lumborum**	Ribs 7–12
Ribs 7–12	**8. Iliocostalis thoracis**	Ribs 1–6, C7
Ribs 3–6	**9. Iliocostalis cervicis**	Transverse processes C4–6
Iliac crest, sacrum, L1–5	**10. Longissimus thoracis**	T1–12, Ribs 4–12
Transverse processes T1–5	**11. Longissimus cervicis**	Transverse processes C2–6
T1–5, C5–7	**12. Longissimus capitis**	Mastoid process
Spinous processes T11–L2	**13. Spinalis thoracis**	Spinous processes T4–8
Nuchal ligament, C7	**14. Spinalis cervicis**	C2
Transverse processes T6–10	**Semispinalis thoracis**	Spinous processes C6–T4
Transverse processes T1–5	**Semispinalis cervicis**	Spinous processes C2–5
C5–8, T1–6	**Semispinalis capitis**	Occiput
Nuchal ligament/T3–T6	**Splenius cervicis**	C1–C3
C7–T3	**Splenius capitis**	Mastoid process/occiput
Ilium/sacrum	**15. Gluteus maximus**	Iliotibial (IT) band/shaft of femur
Ischium	**16. Biceps femoris**	Head of fibula
Upper shaft of femur	**Quadriceps**	Tibial tuberosities
Ischium	**17. Semitendinosus**	Below medial epicondyle of tibia
Ischium	**18. Semimembranosus**	Medial epicondyle of tibia
Condyles of femur	**19. Gastrocnemius**	Calcaneus

The Extensors

Upright posture is a complex process maintained by the action of muscles that extend the head, neck, trunk, and legs. These extensor muscles are located largely on the nape of the neck, the back, the buttocks, and the back of the legs; the only exception is the quadriceps muscles that extend the leg at the knee, which are located on the front of the thigh (Fig. 3-1).

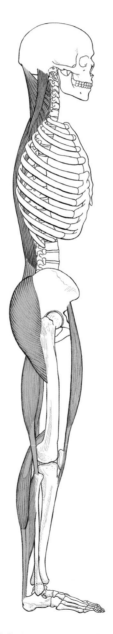

Fig. 3-1. Extensor muscles of back, thigh, and calf regions.

Two key groups of back muscles maintain erect posture. The first group is composed of a series of small muscles running in between the vertebrae of the spine along its entire length, from the sacrum to the occiput of the head. Attaching to the spinous and transverse processes of the vertebrae, these deep muscles act upon the vertebrae, maintaining the internal length and support of the spine (Fig. 3-2).

The second group is the sacrospinalis or erector spinae muscles. In Chapter 5 we look in detail at the deep postural muscles of the spine that comprise the first group, but let's look now at the second group, the sacrospinalis muscles. These are the long muscles that run lengthwise up and down the back, from the sacrum right up to the base of the skull in overlapping bundles that leapfrog from the bottom to the top, forming a continuous sheet of muscles supporting the entire back. When we are standing, lifting weight, bending down, or inclining forward, these muscles maintain the support of the trunk and thus play an essential role in our upright posture (Fig. 3-3).

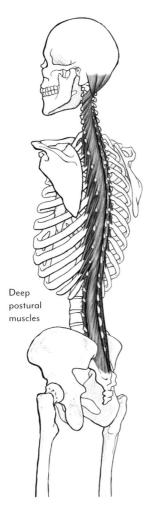

Deep
postural
muscles

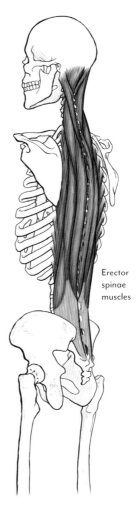

Erector
spinae
muscles

Fig. 3-2. Small postural muscles of the spine.

Fig. 3-3. Sacrospinalis or erector spinae muscles.

Toward the neck, these deeper layers are replaced by more superficial muscles: the semispinalis and splenius muscles. The semispinalis capitis muscle originates at the upper thoracic and cervical vertebra and, passing upward at an oblique angle, attaches to the occiput (Fig. 3-4). The splenius muscle arises from the seventh cervical vertebra and nuchal ligament and passes upward and outward to attach to the mastoid process of the temporal bone and the occiput (Fig. 3-5).

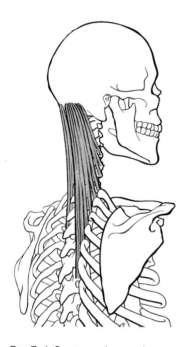 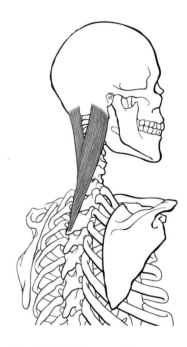

Fig. 3-4. Semispinalis muscles.   *Fig. 3-5. Splenius muscles.*

At the most obvious level, the purpose of the muscles of the neck is to support and move the head. As part of the extensor system on the back of the body, these muscles play a central postural function—a function that becomes even more pronounced during the performance of strenuous activities that require the forceful action of the extensors to support and stabilize the neck and head.

Dynamic Muscle Length and Head and Spine

To support upright posture, the extensor muscles of the spine maintain the support of the trunk and thus play a central role in upright posture. But although these muscles have work to do, they discharge their duty based on the central principle of muscle function described earlier: they must be in a lengthened state and maintain tone in the context of length. Looking at the extensor muscles and where they attach to the skull, you might think that, given how powerful they are, they should pull the head back. In a well-coordinated person, however, this is not what happens. Instead, the head seems to go upward and the muscles of the neck lengthen between their bony attachments (Fig. 3-6). There are two principal reasons for this. The first is that, although the head sits on top of a vertical-placed spine, most of its weight is forward of the point of balance so that it naturally falls forward. Because of this, the muscles of the neck don't simply pull the head back but are kept stretched by the action of the head which, by falling forward, keeps the muscles from shortening. In this way, the muscles are maintained at their optimal length and, most of the time, perform their job of maintaining the balance of the head and spine without actively contracting.

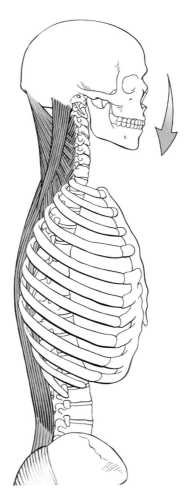

Fig. 3-6. The skull naturally falls forward, creating a dynamic equilibrium in which the muscles of the back are lengthened while maintaining the head in an upright position.

The second reason is that the spine itself, acted upon and supported by a network of muscles and connective tissue, lengthens against gravity, which in turn helps to maintain the length of the long muscles of the back. Because the back muscles lengthen within this bony framework, the entire back feels full and supported, and we can maintain upright posture with a minimum of work and effort. This produces an overall lengthening response of the spine in which the head can be described as going up and the back as lengthening—that is, the head is not pulling back and down but balances forward so that the entire trunk, with the head leading on top, lengthens upward (Fig. 3-7).[1]

In many of us, however, this sheet of muscles becomes shortened and constricted so that, instead of supporting the trunk with lightness and ease, we are in distress much of the time, overworking these muscles and struggling even just to sit. To function properly, this entire sheet of muscles must lengthen in such a way that the head, coming out

Fig. 3-7. Even while using their hands, young children are able to maintain natural upright support with their head balanced forward and their spine lengthening.

of the back, exerts an upward pull upon this sheet so that the muscles, from sacrum to skull, can lengthen and the back can work in a full and supportive way. When we habitually slump, these muscles are disengaged, and the only way to sit upright is to arch the back. This strains the muscles of the lower back, as we quickly learn when we force ourselves to sit upright for a long time and the muscles in the lower back become fatigued and start to burn. This is not how these muscles are meant to work. When the head is balanced forward and the trunk comes to its natural length, the sacrospinalis muscles lengthen between their bony attachments and thus maintain tone in the context of length. In this way, this muscle group functions as a whole to maintain postural support without being strained in any particular region (Fig. 3-8).

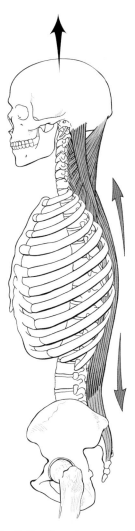

Fig. 3-8. When the postural system is working correctly (as in Fig. 3-7), the balance of the head, together with the lengthening action of the spine, creates effortless support against gravity.

THE UNIVERSAL JOINT

As the uppermost segment of the vertically poised body, the head sits atop the spine and, at this point, can both nod and rotate. Two joints make these movements possible. First, the skull sits on, and articulates with, the first vertebra of the spine—also known as the *atlas* because it holds up the globe of the skull—to form the atlanto-occipital joint. Two rounded bumps on the base of the skull, the occipital condyles, fit nicely into two concave depressions on the atlas, making it possible to nod the head up and down (Fig. 3-9).

Second, the atlas, with the skull sitting on top of it, rotates around the second vertebra—called the *axis* because it forms the pivot, or axis, upon which the first vertebra rotates (Fig. 3-10)—to form the atlanto-axial joint. At the front of the axis is an upward projection called the *odontoid process* (meaning "tooth"), which extends up within the anterior arch of the atlas; the atlas, with the head sitting on it, rotates around the odontoid process, making it possible to rotate the head in relation to the spine.

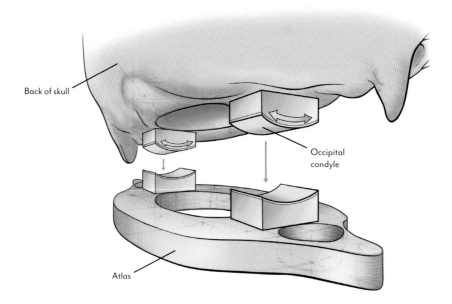

Back of skull

Occipital condyle

Atlas

Fig. 3-9. The skull rocks back and forth on the atlas.

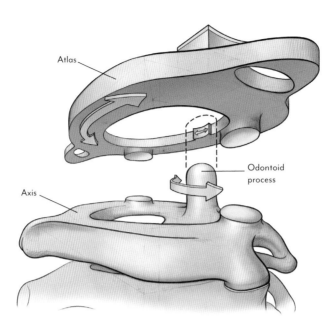

Fig. 3-10. The atlas, with the head sitting on top, rotates around the odontoid process of the axis.

So the skull can nod on the atlas, and it can rotate in relation to the spine because the first vertebra, with the skull sitting on top of it, can pivot on the axis. Together these articulations form a "universal joint" that makes it possible to both swivel and hinge the head (Fig. 3-11).

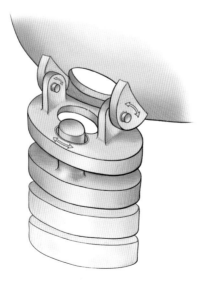

Fig. 3-11. Movement of the head at the atlas and axis makes a universal joint, allowing the head to nod and turn.

HEAD BALANCE

As a spherical object, the head seems to be balanced rather evenly on top of the spine. In fact, the skull is elliptical or egg-shaped and sits on the atlas at a point that is well behind the center of gravity. This means that more of its weight is in front than in back, which causes it to nod or fall forward (Fig. 3-12). Add to this the mass and weight of the jaw, and it becomes clear that the skull is unevenly balanced on the atlas so that its natural tendency is to nod forward at the atlanto-occipital joint (Fig. 3-13).

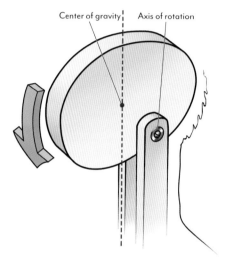

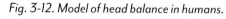

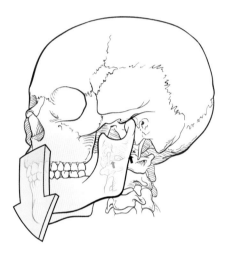

Fig. 3-12. Model of head balance in humans.

Fig. 3-13. The skull (along with the jaw) is weighted forward on the atlas.

THE ANIMAL ORIGINS OF OUR UPRIGHT DESIGN

Although the human upright posture is a unique design found nowhere else in the animal kingdom, it is nevertheless a modified version of the posture of the four-footed animals from which we evolved. In a cat or horse, the head hinges at the front end of the horizontally oriented spine and, in this cantilevered position, exerts stretch on the neck muscles, which in turn maintain the posture of the head (Fig. 3-14). Although radically altered in humans, something of this four-footed arrangement is preserved in the upright human posture, where the head, sitting on top of the spine, sits off balance, and thus continues to exert stretch on the neck muscles. In contrast to the horizontal pull of the cantilevered skull in the four-footed animal, however, the forward balance of the skull on the vertically positioned human spine acts upward against the downward pull of the muscles (Fig. 3-15). This is a very different arrangement than in the four-footed animal. There, the cantilevered head produces a very powerful horizontal pull on the neck muscles which, in turn, maintain tone in the neck to support the weight of the head. In the upright human design, the muscles of the neck and spine pull directly downward, and the forward tilt of the head acts upward on the back of the skull to counteract the downward pulls. This arrangement is more easily disturbed than that of the quadruped because, unlike the animal's head, the human head can easily be pulled backward, thus disrupting the upright support system (Fig. 3-16)—a fact that at least partially explains why humans are so prone to slumping. To counter these downward forces, the head must retain its off-balance position on the spine. The forward balance of the head in relation to the spine is thus an essential feature of the upright human postural system.

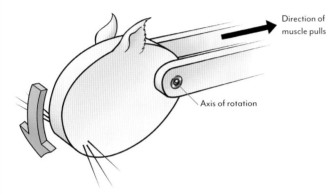

Fig. 3-14. Head balance in a cat.

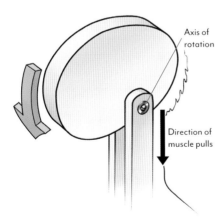

Fig. 3-15. Human head balance.

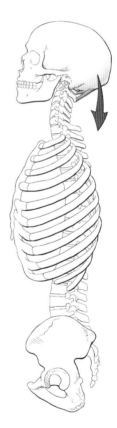

Fig. 3-16. Pulling the head back disturbs its natural forward balance on the spine and interferes with upright posture.

THE PLUMB LINE

We are all familiar with anatomy charts that depict upright posture as a series of parts stacked vertically on the plumb line (Fig. 3-17). This would suggest that, to maintain upright posture, the extensors in back and the flexors in front balance each other to keep us from falling over. But this is not how we are actually designed. We evolved from four-footed animals that, to come upright, had to raise the trunk at the hips. This places the bulk of the demand on the extensors of the back, which have to maintain the support of the trunk. Even when we are standing fully erect, the trunk wants to buckle forward; if we incline forward, bend down, or lift weight, the trunk needs even more support in back (Fig. 3-18). It is the sacrospinalis muscles—as well as the other extensors of the neck, hips, and legs—that bear the brunt of this labor and thus play an essential role in our upright posture.

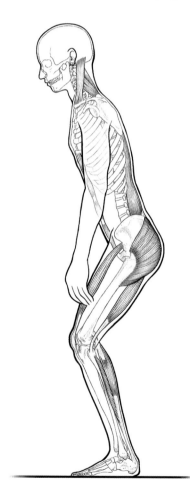

Fig. 3-17. Posture and the plumb line.

Fig. 3-18. When the body is inclined, the action of the extensor muscles, and not the stacking of body parts, supports it against gravity.

EXERCISE:
Semi-Supine Position

To restore length in the extensor muscles of the neck and back, try lying down in the semi-supine position with one or more books supporting your head so that it does not tip backwards, and your knees pointing to the ceiling (Fig. 3-19). It is important, when starting this exercise, to begin by doing nothing at all; this gives muscles that are holding a chance to let go. Then without actively moving, think about your head and knees going away from each other. This will allow your back to lengthen and fill out on the floor. As the muscles of the neck, ribs, and back begin to let go, you will feel that your trunk can begin to regain its natural length, your back can begin to lengthen and fill out on the floor, and your head can begin to come out of the back—a process that will happen automatically as muscles release and naturally tone up.

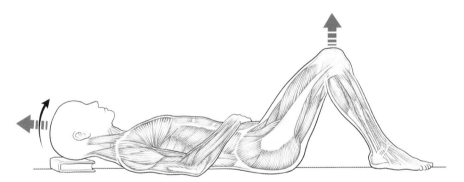

Fig. 3-19. The semi-supine position. By simply thinking of the head and knees going away from each other, the muscles of the back are encouraged to let go into length.

When this toning up happens, the back muscles lengthen and widen, producing the elastic and toned condition of the back that enables it to support the trunk efficiently and effortlessly when you are sitting and standing. This condition cannot be brought about by stretching and working on individual muscles or by positioning the head but instead is a natural condition of the muscles in which they are releasing between their bony contacts so that the entire back has its natural length and breadth and the head balances forward naturally on the spine.

THE SPECIAL ROLE OF THE SUBOCCIPITAL MUSCLES

The suboccipital muscles are a group of four muscles—two oblique and two running more vertically—that link the occiput with the first and second vertebrae, the atlas and axis. These muscles are on both sides of the midline, making a total of eight muscles, or four pairs. There are two suboccipital muscles on the anterior spine, also in pairs, so there are really six paired suboccipital muscles, so called because they are just below the occiput, or base of the skull (Fig. 3-20).

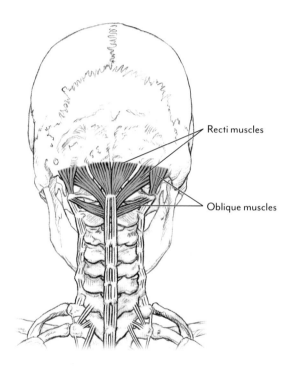

Recti muscles

Oblique muscles

Fig. 3-20. The posterior suboccipital muscles.

The first and most obvious function of the suboccipital muscles is that they move the head—the two recti muscles draw the head backward and the oblique muscles rotate the skull on the atlas. These muscles, which are continuous with the small muscles running the length of the spine, are also postural in function, since they are needed to maintain the support of the head on the spine.

But the suboccipital muscles play another, even more critical role in the human body. As we saw in Chapter 1, embedded in almost all the muscles of the body are small sensory organs called *muscle spindles*, which are designed to detect changes in length as the basis for maintaining muscle tone and stabilizing parts of the body. But there are two things about the spindles in the suboccipital muscles that are different from the spindles found anywhere else in the body. First, more spindles can be found in these muscles than just about anywhere else in the body. And second, though these spindles trigger local tonic reflexes, as other spindles do, they are differentiated by the fact that they play a mediating role in global posture.[2]

The reason for these special characteristics can be found in the unique function of these muscles, which occupy a key role at the top or leading end of the motor system. When we move, we not only have to maintain the stability of specific parts of the body such as the arms or head; we also need to organize the movement or attitude of the body as a whole. If, for instance, you get up from a sitting position, you not only have to incline at the hips and straighten your legs, you also have to move purposefully through space. What organizes this large-scale activity is the relation of the head to the trunk, which "tells" the body what to do as a whole. The suboccipital muscles, as the key muscles of the neck that "sense" the relation of the head to the trunk, help us to do this by organizing body posture in relation to our intention to move or do something (Figs. 3-21 and 3-22).

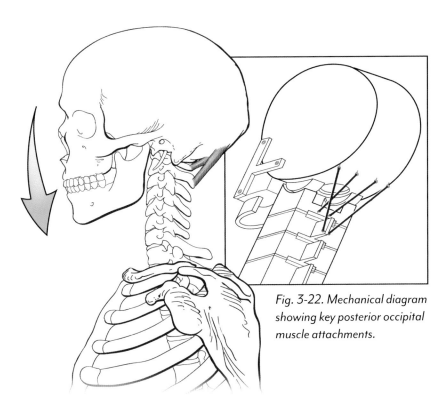

Fig. 3-22. Mechanical diagram showing key posterior occipital muscle attachments.

Fig. 3-21. The suboccipital muscles do not simply move and support the head but are crucial indicators of posture and balance.

The Extensors of the Legs

To maintain postural support against gravity, the legs must be extended at the hips, knees, and ankles. The gluteal muscles are the primary extensors of the legs at the hips; the muscles on the back of the legs, as well as the quadriceps, maintain extension of the knees and ankles when we are standing. Although the legs are powerfully moved by muscles, they also function as extensions of the trunk. In order to perform this function efficiently, the hamstrings must lengthen to the knees (Fig. 3-23).

The hamstrings and gastrocnemius cross the back of the knee (Fig. 3-24). This arrangement helps to stabilize the knee, but if these muscles shorten, the legs become braced at the knee. When the legs and ankles release, the calcaneus releases to the plantar fascia at the bottom of the foot (Fig. 3-25).

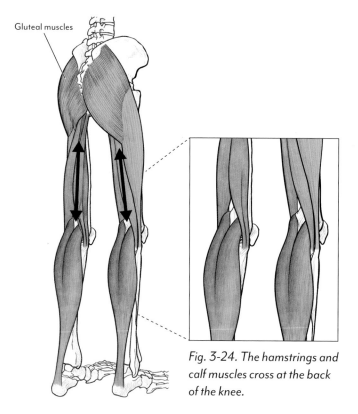

Gluteal muscles

Fig. 3-24. The hamstrings and calf muscles cross at the back of the knee.

Fig. 3-25. The muscles of the calf are continuous with the plantar fascia that runs under the foot.

Fig. 3-23. The anterior muscles of the legs. Paradoxically this view gives the impression that the legs are bent, despite the fact that the leg bones are stacked vertically and are not buckling. Lengthening in the hamstring muscles (black arrows) prevents the knees from being pulled backward.

DYNAMIC MUSCLE LENGTH AND THE LEGS

We've seen that, to maintain upright posture, the neck and back muscles don't simply contract but lengthen between their bony attachments, the head balancing forward and the spine lengthening as a whole. Although the legs are moved by powerful muscles and, in this sense, seem to operate separately from the trunk, they are also part of the postural system, functioning as extensions of the trunk to maintain postural support. This creates a two-way force in which the legs, lengthening out of the lower end of the trunk, create a downward force, and the head, lengthening out of the upper end of the back, creates an upward force. For this to work properly, the leg muscles must be naturally lengthening between their bony contacts and maintaining low-level tone.

Because most adults tend to stiffen and brace the legs, however, the hip and leg muscles become overworked and tight, even during normal standing. Walking and standing are associated, not with a light, balanced support against gravity but with a shortening of stature in which the muscles of the neck, trunk, and legs unnecessarily stiffen and shorten—a disturbance that can be easily seen in the braced attitude of the legs in standing and walking and the stiffened, over-worked quality in the legs when we get out of a chair (Fig. 3-26). To function properly, the legs must maintain length so that the body is supported in a springy and light way and the joints can move freely in bending or walking (Fig. 3-27).

Fig. 3-26. This youth exhibits shortened calf and hamstring muscles, leading to bracing of the legs, pulling backward of the knees, and sinking into the hips.

Fig. 3-27. This little girl is standing with lengthening in the posterior muscles of the leg, with the leg bones stacked vertically.

THE POSTURAL FUNCTION OF THE LEG MUSCLES

When we think of muscles, the first thing that comes to mind is that they contract to produce movement. Muscles are made up of very thin fibers that possess the ability to contract along their length. These fibers are controlled by motor nerves, which activate the chemical activity within the fibers that causes the contraction. All skeletal muscles are made up of bundles of fibers—very few in the case of small muscles and sometimes millions in the case of large ones—that contract to produce movement.

But not all muscle fibers are alike. The large, meaty muscles of the thigh, for instance, are designed to powerfully extend the leg at the knee, whereas the small muscles between the vertebrae are designed to act on the vertebrae and to maintain postural support. Corresponding to their different functions, the fibers making up these different muscles are designed differently. Because the deeper postural muscles of the spine do not perform powerful movements, they have smaller cross sections and contract less forcefully. However, they are also better vascularized and more richly supplied with oxygen, and they fatigue slowly and are, therefore, better suited for long-term work. The muscle fibers found in these regions are called slow-twitch fibers. In contrast, the more powerful thigh muscles have large cross sections and are designed for short, powerful bursts of activity—in some cases, they can produce as much as 100 times the force of slow-twitch muscles—but they fatigue very quickly. The fibers found in these areas are fast-twitch fibers.

We are all familiar with the difference between white and red muscle fibers in the form of light and dark meat in chicken. Chickens fly very little and mostly walk on their two legs. This requires constant, low-level postural support in the leg muscles; accordingly, the muscles of the leg are more richly supplied with blood and are, therefore, dark in color; this is the dark leg meat. The wings are used only occasionally for flying, but when they are used, they must produce quick and powerful bursts of energy. This requires the fast-twitch fibers that function anaerobically—that is, with less supply of blood and oxygen; for this reason, the breast meat, or pectoral muscles, are whiter in color.

In a healthy individual, the deeper, slow-twitch muscle fibers perform the job of maintaining tone and support; the more powerful outer fibers are relatively relaxed and come into play only when vigorous action is required. Research on cats has demonstrated that about half of the muscle fibers making up their skeletal muscles are slow-twitch and half are fast-twitch (Fig. 3-28). The fast-twitch fibers are inactive when they are resting or walking around and come into play only when more strenuous movements are required, which accounts for why the cat's muscles are so soft and pliant. Muscles that are entirely made up of fast-twitch fibers, such as the arm and gastrocnemius muscles, tend to be

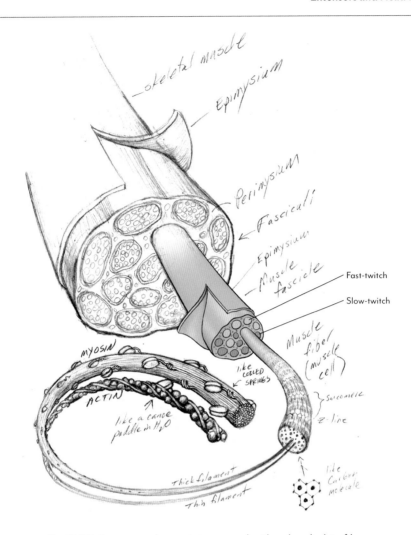

Fig. 3-28. Some muscles are interspersed with red and white fibers.

more on the surface, whereas the deeper postural muscles tend to be comprised entirely of slow-twitch fibers. In muscles that are made up of a mix of fast- and slow-twitch fibers, such as the quadriceps muscle of the thigh, the fast-twitch fibers are located toward the surface, whereas the slow-twitch fibers are located more deeply, where there is more vascular supply. This is why, when we are simply standing, only the deeper slow-twitch postural muscles are required; the powerful fast-twitch portion of the muscle should be relatively inactive and soft and come into play only when we are running or performing other powerful movements, just like the cat.

These two types of muscle differ also in their rates and speeds of firing. *Fast-twitch* muscles are very quick to respond and fire at a quick rate; *slow-twitch* muscles are slower to respond and fire at a much slower rate. The higher firing rate of the white fibers is clearly related to the need for power. The slow rate of the red fibers is designed for the low-level activity of maintaining postural support for long periods. Many muscles, such as the quadriceps, are composed of both types of fiber, but some muscles—the gastrocnemius muscle, for instance—are made up entirely of white fiber. (There is a third type of muscle fiber that is intermediate between fast- and slow-twitch, called *fast fatigue-resistant;* it is not as powerful as the fast-twitch but fatigues less quickly.) All the muscle fibers innervated by a single nerve, or motor unit, are of a similar type—that is, fast-twitch, slow-twitch, or fast fatigue-resistant.

There is a very practical side to understanding the difference between fast- and slow-twitch muscle fibers. When we brace and stiffen the legs, the powerful fast-twitch fibers become overactive and take over the job of maintaining basic postural support. The joints lose mobility and the leg muscles become braced, stiffened, and chronically overworked. These muscles must release so that the deeper muscles can do their job and relieve them from doing the work they are not designed to do. This restores a pliant, natural tone to the leg muscles, relieves joints of strain and restores mobility, and reduces the amount of effort required to stand and walk.

THE GASTROCNEMIUS MUSCLE AND THE FOOT

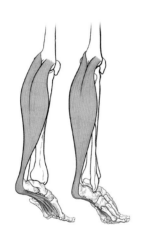

Another example of an imbalance between the fast- and slow-twitch muscles is the overactivity of the gastrocnemius muscle of the calf, which is composed entirely of fast-twitch muscle fibers. Because most of us brace and stiffen the legs, these muscles become overworked. This pulls the heels up so that the feet are not in full contact with the ground (Fig. 3-29), and overrides the action of the deeper postural muscles, which are made up of slow-twitch fibers designed to maintain postural support. For the legs to work properly, these outer fibers must be given a chance to let go and release, allowing the heels to more fully contact the ground and giving the deeper fibers a chance to take on their supporting function. Young children often exhibit this economy of effort, exemplifying in their springy walks and effortless upright support the working of the deeper, slow-twitch fibers. Adults, in contrast, are almost always overworking the fast-twitch fibers of the calf muscles, which become unnecessarily overdeveloped and interfere with the natural supporting function of the leg muscles.

Fig. 3-29. The calf muscles powerfully plantar flex the foot at the ankle.

EXERCISE:
Vertical Stance and the Sacrospinalis Muscles

An easy way to bring about—and to clearly experience—length in the sacrospinalis muscles of the back (Fig. 3-30) is to place yourself in a crouching stance similar to the position used by martial artists. When we are standing fully upright, it is easy to shorten our back muscles, to sink into our hips and to brace our legs. When we place ourselves in the vertical stance, we are organizing the body parts in such a way that they can move away from each other so that, used properly, this position makes it possible to bring about more length in the muscles of the neck, back, and legs, and more freedom in the joints. When assuming the vertical stance, the knees are bent so that the trunk is lowered in space. Bending the knees tends to encourage lengthening of the spine and reduces the curve of the lower back, which helps to lengthen the muscles of the neck and back.

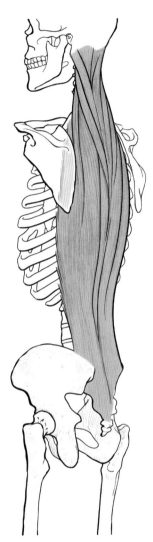

Fig. 3-30. Vertical stance and the sacrospinalis muscles.

Going into the Vertical Stance

1. Place your feet about shoulder width apart so that you can bend easily at your hips and knees. Be aware of your feet on the ground and, without lifting your chest, come up to your full stature (Fig. 3-31a).

2. Bend your knees so that you lower your trunk 5 or 6 inches. Do not flex too much at the knees, which will place pressure on your knee joints (Fig. 3-31b). Also, do not lower yourself slowly but let your knees bend rather quickly.

3. Once you have bent your knees, notice if you have brought your pelvis forward. If you have, this means that you are sinking into your hips and fixing your hip joints. Come up again and try again; this time see if you can prevent your hips from going forward as you bend your knees.

4. You should now be balanced nicely over your feet with your knees tending to go away from your hips, your hips staying back from your knees, and your trunk lengthened.

Directing in the Vertical Stance

Once you are in a balanced position, you are ready to give the following directions, or thoughts:

1. Think of allowing your head to go forward and up and your pelvis to drop away from your head so that you are allowing your back to lengthen (Fig. 3-31b). Do not tuck your pelvis but simply allow the head to go forward and up and your pelvis to go away from your head.

2. You should now feel the length of your back from the back of your head right down to your sacrum. This means that you are allowing the muscles of the lower back to release and that your back and neck muscles are tending to lengthen. Performing this movement isn't simply about posture or body mechanics; it is intended to bring about muscle length through the opposition of body parts as the basic principle governing body support.

Neither of the directions just enumerated are voluntary movements but are instead intended to happen simply by thinking about them and allowing muscular release to take place on its own. If, for instance, you think of the knees going away from the hips by actually moving them forward, the hips will move with the knees and there will be no release of muscle tension, just an overall movement of the legs. Do nothing and, instead, simply think of the knees going away from the hips; this allows the thigh muscles to release and brings about more lengthening in the muscles of the legs.

Also, do not lose one direction when giving another. It is easy, for instance, to think first of allowing the knee to go forward and, a moment later, to let the hip move back. Be sure not to move body parts and, instead, allow release to happen internally by thinking of letting go between the two points.

Vertical Stance with Added Directions

When you have tried this exercise a few times, try doing the vertical stance with the following added instructions:

1. Bend your knees and, without changing position, think of releasing your pelvis to drop away from your head and your head to move away from your pelvis so that your back can lengthen.

2. Think of releasing your calves by letting your heels go onto the ground and letting your knees go away from your heels. This is not a movement; simply think of allowing the muscles to lengthen or release between the heels and the back of the knee. Continue to allow your pelvis to drop away from your head and your head to go up and forward by not tightening in the back of your neck. With your knees releasing away from your heels, you may feel some extra release in your back so that you can feel the length of your back from sacrum to skull (Fig. 3-31c) and greater elasticity in the muscles of the legs, buttocks, back, and neck.

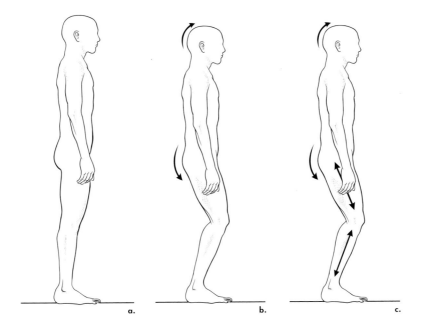

a.   b.   c.

Fig. 3-31. The vertical stance exercise.

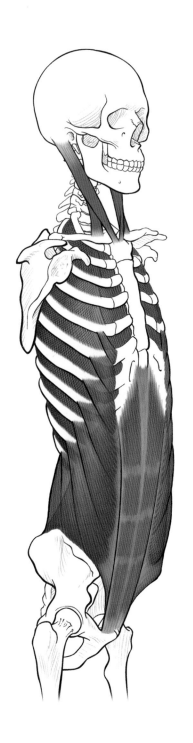
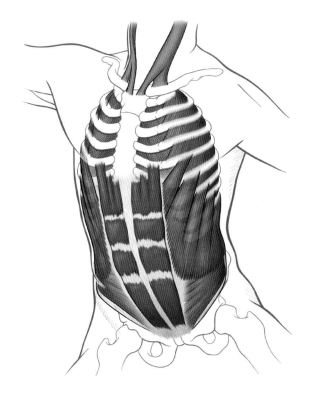

Flexors and Front Length

4

The *flexors* form an extensive sheet of muscles on the front of the body that includes the abdominal muscles, the intercostals, and the sternocleidomastoid muscles. The major functions of the flexor muscles are to balance the action of the extensors and to flex the trunk, but just as critically, these muscles act as a tensile structure which, suspended between pelvis and skull, supports the abdominal contents and rib cage and thus indirectly helps to maintain the length of the trunk.

Key to the Flexors

Although the flexors are capable of powerfully flexing the trunk, the key to the flexors is not their strength but their ability to provide tensile support for the trunk so that, instead of dragging down *upon* the skull, the rib cage and innards are freely suspended *from* the skull. Because the flexors ultimately attach to the base of the skull, their length depends, as do the extensors in back, on the forward balance of the head on the spine.

Contraindications

If the flexors are shortened and drag down upon the skull, the system loses length and cannot maintain upright support. When this happens, the back muscles disengage and, when we need to maintain upright support, we overuse the lower back muscles and raise the chest to compensate, maintaining our upright support in a way that, paradoxically, prevents the natural support system from working.

WHEN WE LOSE FRONT LENGTH

1. The sternocleidomastoid muscles shorten and pull the head back.

2. The upper spinal column is dragged downward.

3. The rib cage and abdominal region are shortened and collapsed.

WHEN WE REGAIN FRONT LENGTH

1. The intercostal muscles, rib cage, and chest open up.

2. The upper spinal column regains length.

3. The head balances forward and up on the atlas.

4. The back lengthens and widens.

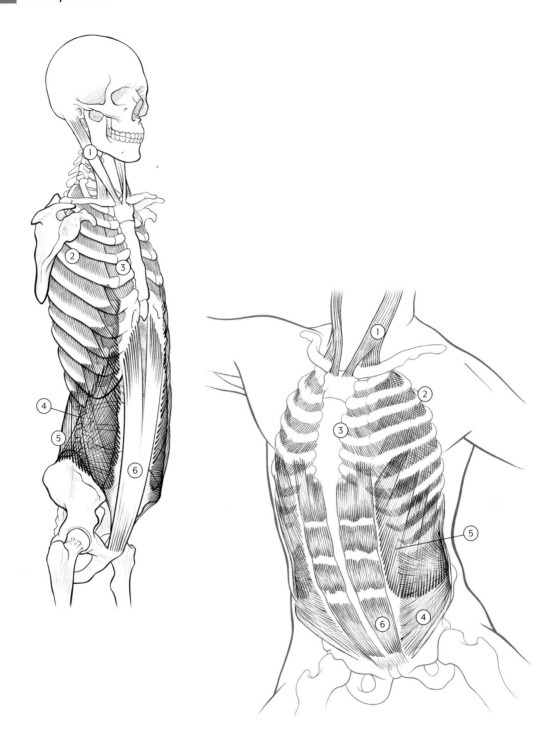

ORIGIN	MUSCLE	INSERTION
Sternum/clavicle	**1. Sternocleidomastoid**	Mastoid process
Lower border of rib	**2. External intercostal**	Upper border of adjacent rib
Lower border of rib	**3. Internal intercostal**	Upper border of adjacent rib
Iliac crest/lumbar fascia	**4. Internal abdominal oblique**	Pubic bone/costal arch, Ribs 1–12
Ribs 5–12	**5. External abdominal oblique**	Iliac crest
Lower sternum	**6. Rectus abdominis**	Pubic bone

The Flexors

In a very general sense, the muscles on the front of the body flex the neck and trunk and thus act antagonistically to the extensors in back. The extensors of the back maintain the upright support of the trunk and hyperextend the trunk; the muscles in front counterbalance the extensors and actively flex the trunk, as when we do sit-ups (Fig. 4-1).

Yet such descriptions hardly do justice to the role of the flexors in the front of the body, or to their importance to the proper working of the musculoskeletal system as a whole. If we were built like upright columns, we would require muscles to stabilize the column both in front and in back, and the two sides would be equal. But this is not how we are constructed. In a four-footed animal, the internal organs and rib cage hang below the spine, which acts as a bridge between the fore and hind limbs (Fig. 4-2). To come upright, the ribs and internal organs below the body now hang in front (Fig. 4-3).

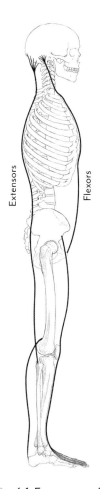

Fig. 4-1. Extensors and flexors.

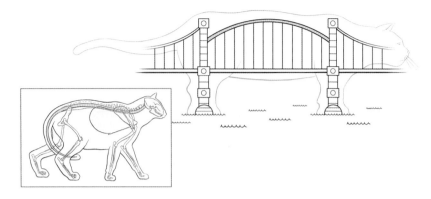

Fig. 4-2. In a four-footed animal, the spine forms a bridge between fore and hind limbs and the internal organs hang below the spine.

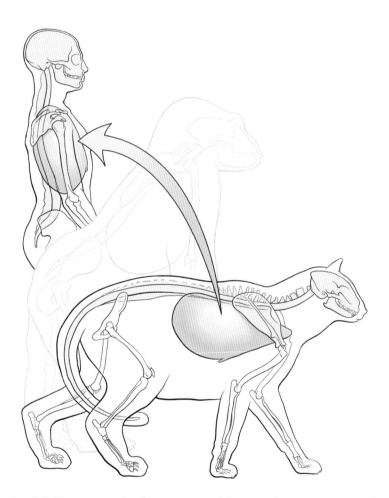

Fig. 4-3. To come upright, the orientation of the spine changes in space so that structures that are below the spine in four-footed mammals are in front of the spine in humans.

This gives the two muscle systems very different roles. The extensors, as we saw, carry the bulk of the duty of supporting the trunk at the hips and legs. The flexors in front do little to directly support the trunk and do not even attach directly to the spine, functioning as a tensile structure for supporting the abdomen and rib cage in front (Fig. 4-4). If these structures drag down in front, the system loses length and cannot maintain upright support, which is what happens when we spend a great deal of time being sedentary and slumping. When this happens, the back muscles disengage and, when we need to maintain upright support, we overuse the lower back muscles and raise the chest to compensate, maintaining our upright support in a way that interferes with the natural support system.

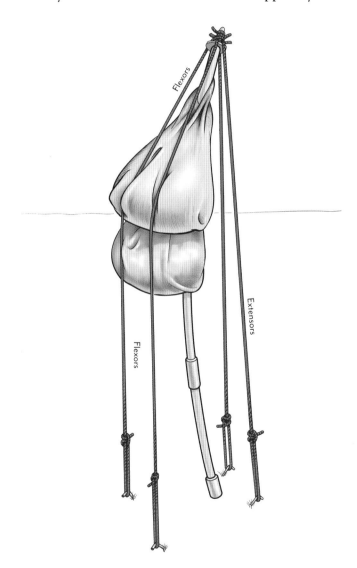

Fig. 4-4. The flexors function as a tensile structure for supporting the weight in front.

The Flexor Line

The most obvious line of muscles on the front of the body is made up of the abdominal and sternocleidomastoid muscles, which form a continuous line of support from the pubic bone to the sternum, and from the sternum right up to the base of the skull where the sternocleidomastoid muscles attach to the mastoid processes (Fig. 4-5). The front of the body, then, is suspended from the skull; if we want to maintain length on the front of the body, we have to make sure that we don't let this system drag down upon the skull, which is what happens if we are sedentary and collapsed, in which case we collapse the chest so that the entire front of the body loses support and we go into a slump (Fig. 4-6).

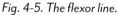

Fig. 4-5. The flexor line.

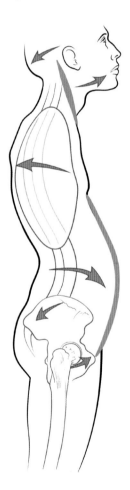

Fig. 4-6. When we slump, we disengage the extensor muscles in back so that the spinal curves are accentuated, the head is pulled back, and the pelvis rotates forward. This results in the pulling down of the trunk and collapse in front.

The Flexors and Head Balance

The sternocleidomastoid muscle attaches to the base of the skull at a very important point called the *mastoid process*, which is the large bump just underneath your ear lobe on either side of the skull (Fig. 4-7). This bump is part of the temporal bone, which is situated on the sides (and underside) of the skull. It provides a point of attachment for the sternocleidomastoid muscles, which originate at the sternum and clavicle and angle up and backward to attach to this point on either side of the skull. What is so important about this attachment is that, because the mastoid process is just behind the point of balance of the skull, the sternocleidomastoid muscle pulls the head not forward, but backward in relation to the spine (Fig. 4-8). This means that although these muscles tend to pull the spinal column forward and drag the skull downward, they pull the skull itself, where it balances at the atlanto-occipital joint, backward. This is something you can clearly see if you slump in a chair: the head and neck are dragged forward, but the head is pulled back in relation to the spine.

But why should these muscles attach to the skull at a point that pulls the head back and not forward at the atlanto-occipital joint? The answer is that, just as the head balances forward and up in order to counteract the pull of the extensors in back, the head must go forward and up in order to counteract the pull of the flexors

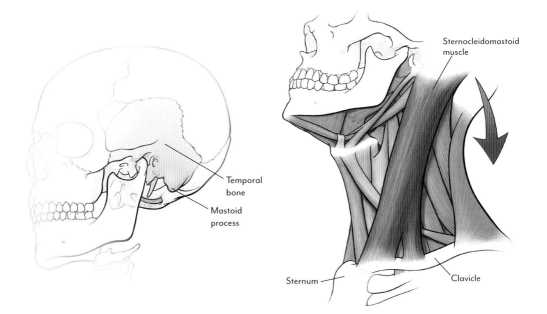

Fig. 4-7. Skull showing the mastoid process.

Fig. 4-8. When the extensors disengage leading to shortening in front, the action of the sterno-cleidomastoid muscle pulls the head backward in relation to the spine.

in front. In this way, the balance of the head helps to maintain the length of the body in front, and the rib cage, instead of dragging down upon the head, is suspended from the head (Fig. 4-9).

We have, then, two basic systems of muscles—the extensors in back and the flexors in front—both of which attach to the skull, which is balanced forward at the atlanto-occipital joint. In order to maintain upright posture, these systems have to maintain length, and they do so in relation to the oppositional forces of skeletal parts, including the skull, which acts as a kind of counterbalance on top of the entire structure that helps to maintain the length of the body in front and in back. When we have our full length in front, the head goes forward and up and, in this context, its forward-and-up direction is part of maintaining the length of the trunk in front. The head, and its forward-and-up direction in space, is part of this dynamic system of muscles counterbalancing the action of these muscle groups in such a way that the trunk lengthens easily and effortlessly against gravity.

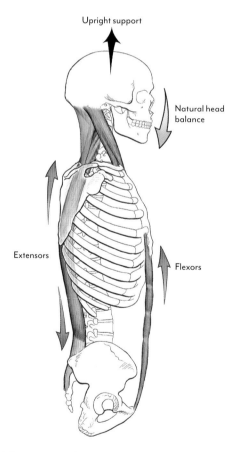

Fig. 4-9. When the flexors and extensors of the trunk lengthen, the head naturally balances forward. This results in natural upright support against gravity.

THE STERNOCLEIDOMASTOID
MUSCLE AND ITS ATTACHMENT

The sternocleidomastoid muscle, as we just saw, attaches to the mastoid process, which is part of the temporal bone. Although the mastoid process is quite close to the point of balance on the skull, the muscular attachment of the sternocleidomastoid is decidedly biased toward the back and so has the general effect of pulling the head back at the atlanto-occipital joint (Fig. 4-10).

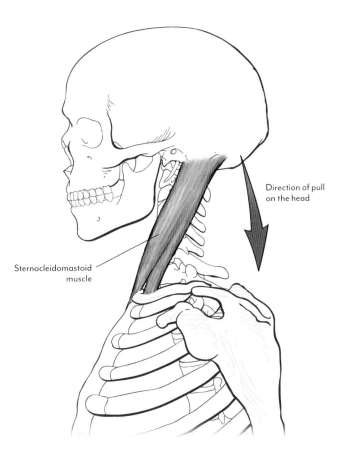

Direction of pull
on the head

Sternocleidomastoid
muscle

Fig. 4-10. If the only muscle supporting the skull were the sternocleidomastoid, the resulting effect would be to pull the head backward. This can happen when we slump, thus disengaging the extensor muscles in back.

THE MUSCLES THAT SUPPORT UPRIGHT POSTURE

It is quite natural to assume that, in order to stand upright, we need muscles on both the front and the back of the body. We can, after all, fall backward and forward, so it makes sense that we should have muscles to stabilize us on both sides; and just as we need to be able to extend the body backward, we need to be able to flex it forward, as we do when we are lying on our back and do a sit-up to raise the trunk. The misleading idea that these muscle groups balance each other is further strengthened by the notion that muscles act antagonistically, which is a big reason why, if we are having back trouble, we are urged to "strengthen the core"—the muscles on the front of the body—in order to balance the action of the muscles in front and the muscles in back.

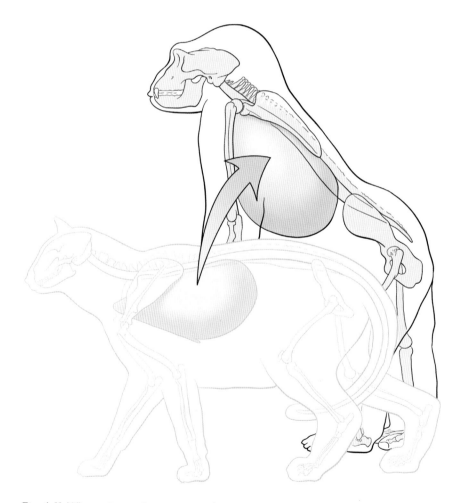

Fig. 4-11. When primates began to stand up on their hind limbs and raise the trunk to a more vertical position, the rib cage and organs that hung below the spine were now out in front of it, placing the bulk of the work on the extensors in back.

All of this, however, assumes that we were somehow created in the upright pos-ture, and that maintaining balance on two feet consists of placing muscles that can act equally upon both sides of the body. It also assumes that good posture consists of organizing body parts in a vertical arrangement, as if we are made up of blocks stacked one on top of the other. Nothing could be further from the truth. If you observe a four-footed animal, you see that the spine acts as a kind of bridge between the fore and hind limbs, and that the rib cage and innards hang below the spine. When the first primates stood upright on their hind limbs and raised the trunk to a more vertical position, the rib cage and organs that hung below the spine were now in front (Fig. 4-11). What this means is that, to maintain upright posture, the main burden falls not on the flexors in front but on the muscles in back, which have to keep the head and trunk from falling forward and the trunk extended at the hips. In other words, when it comes to upright support on two feet, not all muscles—or muscle groups—are equal (Fig. 4-12).

Fig. 4-12. In order to maintain upright posture, we need to prevent our joints from buckling—the main task for the four men here, who take on the role of the extensors in keeping us erect.

What then do the front muscles do? Not only do they have less work to do than the extensors, but they do not even act directly on the spine, since they wrap around the abdomen and rib cage. The answer is that, in addition to stabilizing the front of the body, they maintain the lengthened support of the trunk. With the abdomen and rib cage hanging in front, this is not a simple task, as we can see when we begin to slump and lose length in the front of the body. Slumping in this way is more than bad posture; it means that the muscular system is no longer working effortlessly to support us against gravity, forcing us to compensate by overusing the extensors in back. In order to support ourselves, the back has work to do, but the front also plays a crucial role, which is to maintain the lengthened support of the structures in front.

THE ANATOMY OF THE FLEXOR MUSCLES

We've seen that, in four-footed animals, the rib cage and viscera hang below the spine; in humans, the rib cage and viscera hang in front. In practical terms this means that, in contrast to the extensors, which lie close to the spine and act directly upon the trunk, the flexor muscles do not hug the spine but, wrapping around the gut and rib cage, are well removed from the spine (Fig. 4-13). That is why the flexors do not form a sheet of muscles in front mirroring those in back but encircle the cylinder of the trunk (Fig. 4-14).

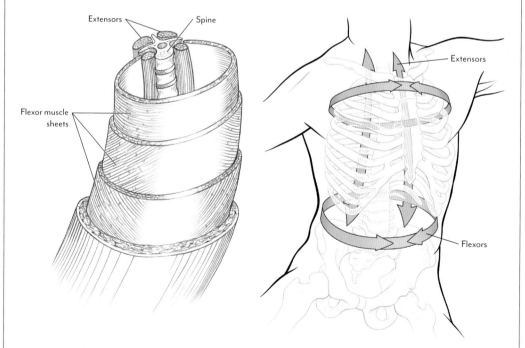

Fig. 4-13. Unlike the extensors, the flexors do not hug the spine but wrap around the trunk.

Fig. 4-14. Flexors wrapping cylindrically around the trunk.

EXERCISE:
The Importance of Front Length

If the body drags down in front, the system as a whole loses length and cannot maintain upright support, which is what we see today even in young children who spend a great deal of time slumping heavily while sitting in front of computers and handheld devices. When this happens, the back muscles disengage and, when we need to maintain upright support, we overcompensate by raising the chest and tightening the lower back muscles, maintaining upright support in a way that, paradoxically, prevents our natural support system from working properly. To overcome this tendency, it is necessary to restore natural length by giving the muscles in the neck and back a chance to lengthen while, at the same, allowing the front to open up. If we do this in a crude way—that is, by simply lifting up the chest and arching the back—this doesn't engage the natural support system but further interferes. This is why it is so important to restore front length in a way that allows the back to lengthen as well.

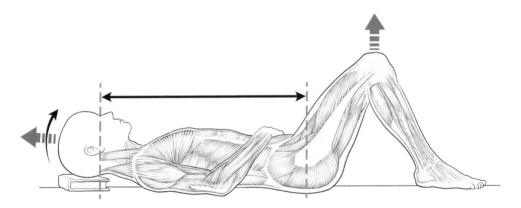

Fig. 4-15. The semi-supine position and regaining front length.

The semi-supine position discussed on page 59 is ideal for restoring front length because it gives the front of the body a chance to open up while the back muscles, instead of over-arching, can lengthen and fill out as well. One way to think about restoring front length is to release the neck muscles to let the back lengthen and widen and the knees to go away, while at the same time allowing the front to open up from the pubic bone right up to the base of your skull. This exercise is not about "doing" but about "thinking," or what is sometimes referred to as "directing." For more information on the nature of directing and how to direct, see the appendix on page 287.

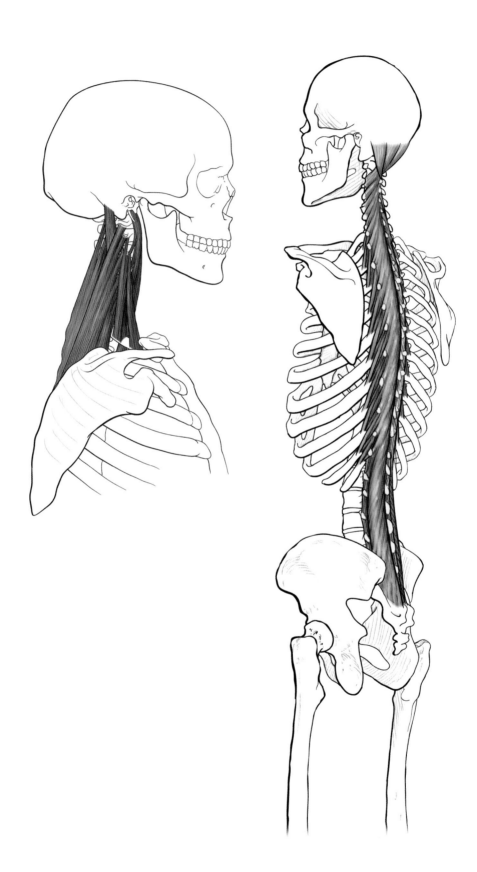

Lengthened Support and the Healthy Spine

5

The spine, or backbone, is the central bony structure upon which muscles act to support us against gravity and to produce movement. Many exercise and relaxation systems emphasize the role of muscles in musculoskeletal health through focusing on stretching, relaxing, treating, and strengthening muscles. But it is the combination of muscles, connective tissue, and bones working synergistically together that produces upright support. The spine is the central structure around which this muscle/connective tissue network is organized to create a balanced, supportive whole.

Length: The Key to the Spine

As the central bony structure supporting the neck and trunk, the spine forms a compression-resistant column for bearing the weight of the skull and the trunk as a whole. To function properly, however, muscles must act upon the spine so that, instead of collapsing, it actually lengthens against gravity.

Contraindications

The universal tendency, in our current sedentary way of life, is to collapse the spine, exaggerating the thoracic and cervical curves. In this state, the deep muscles of the spine are inactive, sometimes even becoming atrophied.[1] At the same time, the lumbar muscles are shortened and contracted because, to counter this state of collapse, they must compensate by overworking, which further shortens the spine.

WHEN THE SPINE IS SHORTENED

1. The spine collapses and loses it natural curves.

2. The head is pulled back and there is a loss of front length.

3. The deep muscles of the spine are inactive.

4. There is increased pressure on the vertebrae and intervertebral discs.

WHEN THE SPINE IS LENGTHENING

1. It is possible to sit fully upright with a minimum of effort and strain.

2. The head balances forward and up and the pelvis drops away from the head.

3. The deep muscles of the spine are toned and active.

4. There is minimal pressure and strain on the vertebrae and intervertebral discs.

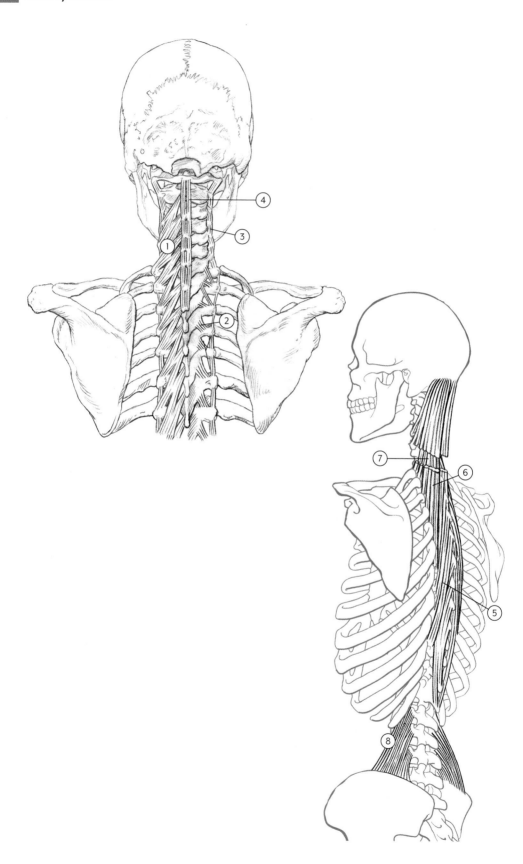

ORIGIN	MUSCLE	INSERTION
Sacrum/transverse process	1. Multifidus	Spinous process of adjacent vertebrae
Transverse process	2. Rotatores	Lamina of adjacent vertebrae
Transverse process	3. Intertransverse	Transverse process of adjacent vertebrae
Spinous processes	4. Interspinales	Spinous process of adjacent vertebrae
Transverse processes T6–10	5. Semispinalis thoracis	Spinous processes C6–T4
Transverse processes T1–5	6. Semispinalis cervicis	Spinous processes C2–5
C7–T6	7. Semispinalis capitis	Occiput
Iliac crest	8. Quadratus lumborum	L1–5, T1

The Spine

The spine, or backbone, is a flexible column formed by the vertebrae and the intervertebral discs. There are thirty-three vertebrae in all: seven in the cervical (neck) region, twelve in the thoracic (chest) region, five in the lumbar region, five in the sacral (pelvic) region, and four in the coccygeal (tailbone) region. The sacral and coccygeal vertebrae are fused together to form the sacrum and coccyx, so for all intents and purposes the spine consists of twenty-four moveable vertebrae, plus the sacrum and tailbone (Fig. 5-1).

The main purpose of the spine is to bear weight and to provide a support structure for muscles to act upon in producing upright balance and movement. Each vertebra has two parts. The front part, called the *body* or *centrum*, is round and forms a weight-bearing column of bony segments stacked on top of each other with intervertebral discs in between (Fig. 5-2a and b). The *intervertebral discs* are resilient and pliable structures designed to absorb shock and to allow twisting and bending of the spine; together with the vertebral bodies, they form a strong yet flexible column for supporting the head and trunk. The vertebral bodies and the intervertebral discs are smaller at the top and get much larger and stronger at the bottom, where they must support greater weight.

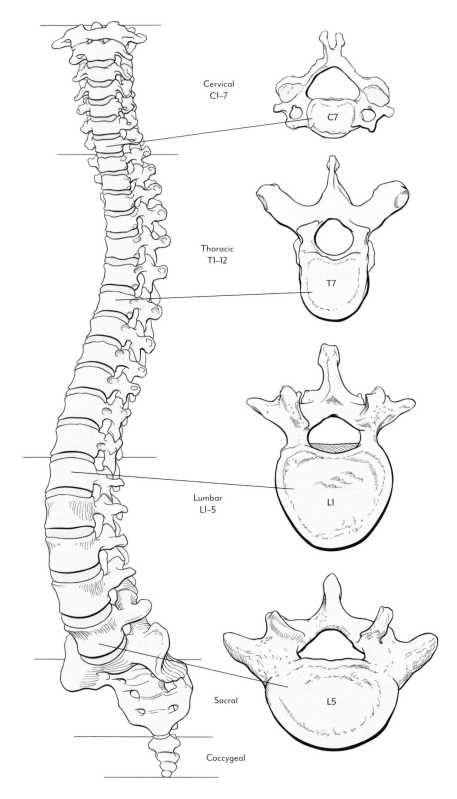

Fig. 5-1. The spine consists of twenty-four movable vertebrae, plus the sacrum and coccyx.

The back part of the vertebrae is the *vertebral arch,* which protects the spinal cord and also provides attachments for ligaments and muscles. These ligaments bind the vertebrae together and limit the range of motion of the vertebrae and intervertebral joints (Figs. 5-3 and 5-4). There are also surfaces that articulate with the vertebrae above and below, forming a series of joints (Fig. 5-2c and d).

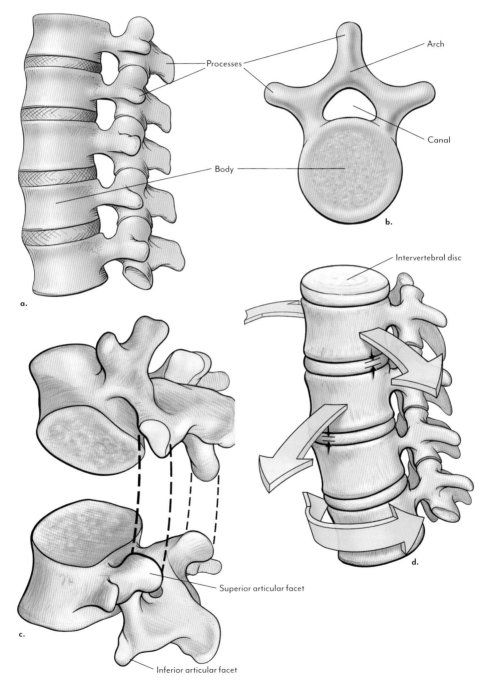

Fig. 5-2. Vertebrae of the spine: a. Vertebral column; b. Basic structure of a typical vertebra; c. Vertebrae stack vertically with interlocking superior and inferior facets creating a movable joint; d. The pliable intervertebral discs allow movement of the vertebrae in all three planes.

Stacked thoracic vertebrae

Ribs

Fig. 5-3. The ligaments of the spine form a latticework that stabilizes the vertebrae and the ribs that attach (as here) in the thoracic spine, while at the same time allowing movement in three dimensions.

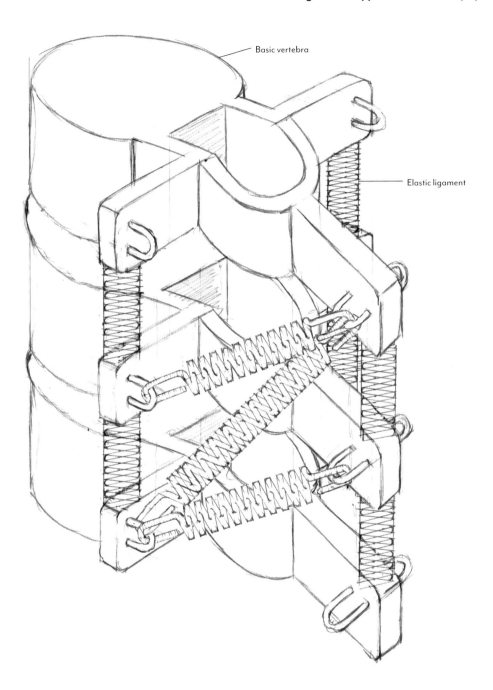

Basic vertebra

Elastic ligament

Fig. 5-4. This mechanical drawing demonstrates the elastic support that the ligaments provide to the spine, while at the same time maintaining the fidelity of its structure.

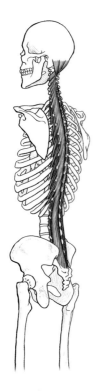

Fig. 5-5. The spine is supported by a series of intervertebral muscles.

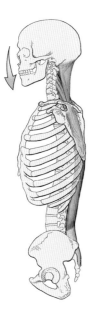

Fig. 5-6. When the muscles of the neck and back are functioning properly, the spine lengthens.

With only ligaments supporting it, however, the spine is a rather inert structure that would collapse under the weight of the trunk. It is the intervertebral muscles, acting upon the vertebral attachments, that support the curves of the spine and thus maintain its internal length. Thus the spine acts as a supporting column but is itself supported by a series of small postural muscles that maintain its internal support, aided by muscles on the anterior spine in the lumbar and cervical regions (Fig. 5-5).

But these intervertebral muscles cannot perform their function simply by "pulling" on the vertebra. When the spine is collapsed, as in the familiar postural slump, the postural muscles are disengaged and the spine has no internal support. As we have seen in previous chapters, for the postural muscles to function in their supporting role, the neck muscles must be lengthened and the head must balance forward so that the spine can regain its natural length. This allows the deeper postural muscles to come into play and to act upon the vertebral processes. The spine then acts as a lengthening device in which muscles act upon bones, but bones oppose each other so that the muscles perform their function within this larger context (Fig. 5-6).

When the spine is supported in this way, we can see that, instead of being compressed by the head and trunk, it lengthens as a whole and the intervertebral discs, which regain their natural buoyancy, exert a hydraulic force against downward pressure. In this sense, the human spine, in addition to supporting weight and providing attachments for muscles to act upon to produce movement, acts as a lengthening device with hydraulic as well as compression-resistant properties. In this dynamic state, the spine does not function simply as a rigid compression-resistant column upon which muscles act. The muscles and spine work together to form a complex tensegrity structure that is able to dynamically lengthen in response to gravitational forces (Fig. 5-7). For this system to work properly, muscles do not simply pull on the spine but also lengthen to allow the spine to assume its full length. Thus we see that head balance in relation to the lengthening spine/muscle complex acts as the central mechanism for maintaining active and dynamic lengthening support against gravity.

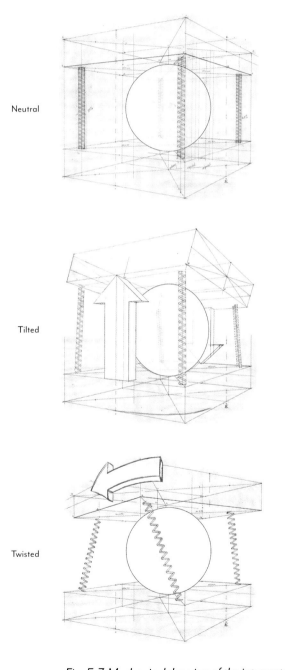

Neutral

Tilted

Twisted

The intervertebral discs are firm but pliable. The support they give can be likened to an inflatable ball sitting between two planks of wood. The discs allow each vertebra a considerable range of movement, which, as we saw previously, is checked by the ligamentous binding, and the vertebra's own bony outcrops that act like brakes to prevent excessive strain and eventual rupture.

Fig. 5-7. Mechanical drawing of the intervertebral discs.

The Spinal Curves

The four curves of the human spine are an essential part of our upright support system. A four-footed animal has only two primary curves: the broad thoracic curve of the spine between fore and hind limbs, and the cervical curve supporting the head (Fig. 5-8). In the upright posture, a lumbar curve is required to counterbalance the thoracic curve (Fig. 5-9).

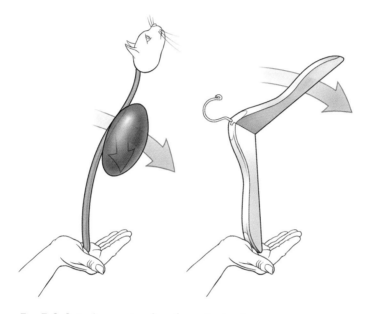

Fig. 5-8. Spinal curves in a four-footed animal.

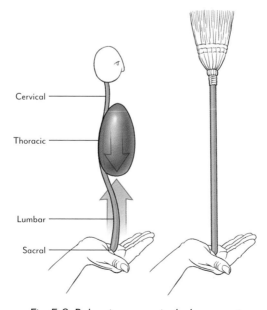

Fig. 5-9. Balancing curves in the human spine.

THE SPINE/CONNECTIVE TISSUE COMPLEX

As the body's central bony structure, the spine is often depicted as a sturdy column upon which muscles act to maintain upright posture. But muscles do not simply act on bones, and the spine is not simply a column for supporting the weight of the head and trunk. To function as a supporting column, vertebrae, connective tissues, and muscles work together to create tensegrity support. The spine is nothing without its muscle and connective tissues, all of which act in partnership to produce dynamic, lengthened support (Fig. 5-10).

This partnership between bone, connective tissue, and muscle, working as a dynamic whole, is essential to what it means to have a healthy spine. If the head and spine are working oppositionally, and if the small muscles of the spine act upon these bony attachments to provide postural support, the spine is lengthened and is not subjected to unnecessary stresses. If, on the other hand, the muscles and bones are not acting in cooperation to provide this support, we are forced to slump, which places undue stress on the intervertebral discs and ligaments (Fig. 5-11a). Muscles become flaccid and weakened and, even if the spine is passively aligned or stretched, it will again go into imbalance. Recognizing this, many of us try to strengthen or align the spine in the hope of establishing proper postural support (Fig. 5-11b). But the muscles do not work separately from the vertebrae and, once we return to normal activity, the spine goes back into collapse and misalignment. We can try to improve spinal health by strengthening the back and core muscles that support the spine, but this cannot activate the deeper postural

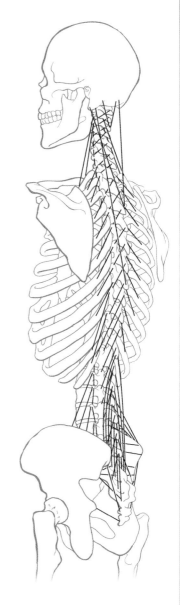

Fig. 5-10. The spine is a tensegrity structure.

muscles that support the vertebrae. Strengthening the outer muscles, in fact, will often override the action of the deeper postural muscles, which cannot function if the larger muscles are overactive and placing stresses on the spine that prevent it from lengthening.

 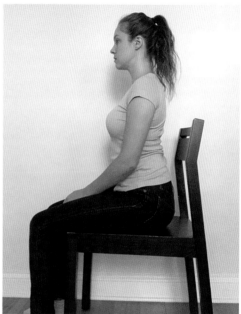

Fig. 5-11. a. When we slump, the postural muscles of the spine are disengaged; b. When we deliberately sit up, pulling the back upright, we overwork the muscles of the lower back.

For the spine to function properly, all its parts must be active and working in cooperation, distributing forces throughout so that all the parts are in balance. In order for this to happen, the head must balance forward and the trunk must lengthen so that the head and pelvis oppose each other—in other words, the key parts of the upright system must work oppositionally so that the spine is naturally lengthened and the deeper muscles are able to function in their supportive role (Fig. 5-12). No exercise or treatment can achieve this condition, which is part of its natural design and can be brought into being only by understanding how the spine naturally lengthens

Fig. 5-12. When the muscles and spine work together, the system lengthens dynamically.

when its constituent parts are working in accordance with this design. In short, the spine can be healthy only when it functions dynamically in the context of its larger design.

ORIGINS OF THE SPINE:
THE NOTOCHORD

The spine finds its origins in the notochord, a flexible beam found in the dorsal area of primitive chordates—the precursors of fish—such a lampreys. This beam enables the animal to move through water by flexing its body from side to side. The notochord evolved to develop a bony surround—the vertebral arch—that protected the spinal cord running above it, and the chord itself ossified to form the vertebral bodies and discs, all of which are characteristic of vertebrates today (Fig. 5-13). In contrast to the rigid exoskeletons of insects, this internal bony framework provided a useful template for vertebrates that came out of the water, allowing for a great variety of designs for moving efficiently on land. In four-footed animals, the spine acts as a bridge between the fore and hind limbs (Fig. 5-14).

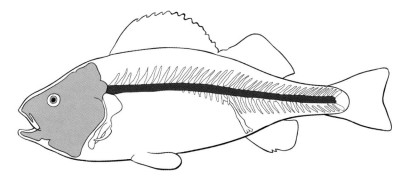

Fig. 5-13. The spine of a fish.

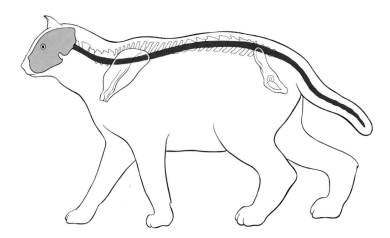

Fig. 5-14. The spine is a bridge in four-footed animals.

To support upright posture, the human spine requires an extra curve in the lumbar region. The spine of a four-footed animal forms a broad curve like the arch of a bridge, with the fore and hind limbs at either end. This design is not well suited for upright posture because, when the trunk is raised up, there is too much weight in front. Primates that stood up on two legs began to develop a lumbar curve that made it possible to balance the trunk upright on two legs. In this way, the human spine developed balanced curves that enabled it to function as a vertical support structure (Fig. 5-15).

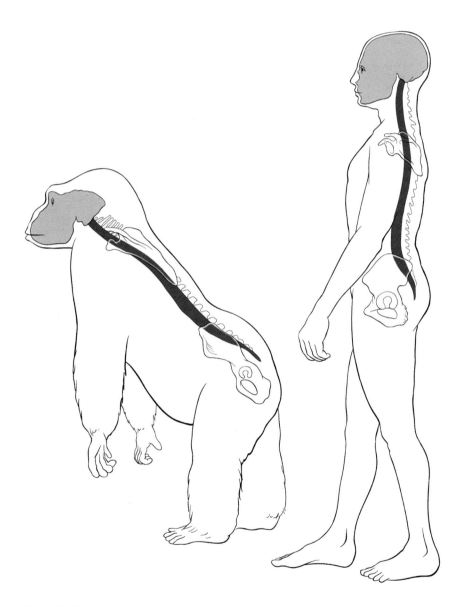

Fig. 5-15. The lumbar curve started to develop in primates, becoming fully formed in upright humans.

ROTATION AROUND A VERTICAL AXIS

Our upright bipedal posture makes it possible to rotate the head and trunk around a vertical axis, giving us a 360-degree field of vision and the ability to look around, even while walking—a unique feature of our human design (Fig. 5-16).

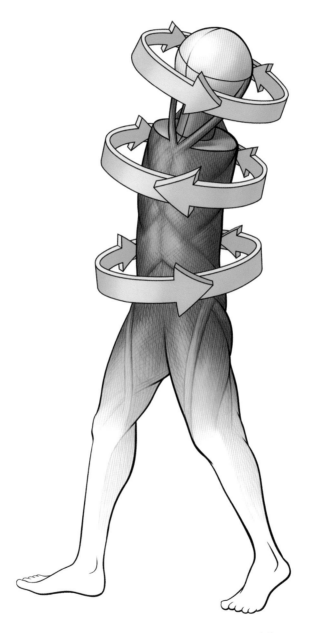

Fig. 5-16. Upright support permits rotation around the vertical axis.

THE HUMAN SPINE IS A LENGTHENING DEVICE

In humans, the spine functions as a weight-bearing column that transfers the weight of the trunk through the pelvis and onto the heads of the femurs (Fig. 5-17). But the spine also acts as a lengthening device that, in sitting and standing, counteracts the pull of gravity (Fig. 5-18).

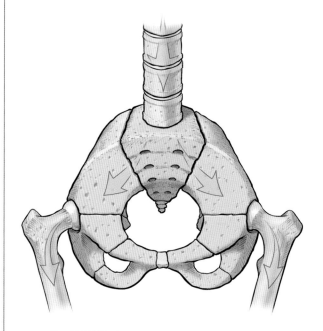

Fig. 5-17. The human spine is a weight-bearing column.

Fig. 5-18. The spine functions as a lengthening device.

THE SPINE: WEIGHT-BEARING COLUMN OR TENSEGRITY STRUCTURE?

As the uppermost portion of the spinal column, the cervical spine is a compression-resistant structure designed to support the weight of the head. But the neck vertebrae would be sorely tested if the weight of the head were entirely supported on so small a base as the cervical column. In fact, the upper spinal column is supported by a network of ligaments, muscles, and connective tissues that form a complex rigging, like the shrouds on a ship's mast. Working with the spine, this rigging forms a tensegrity structure that distributes the forces acting on the spine to the muscles, connective tissue, and ligaments (Fig. 5-19). For this reason, the actual pressure of the head on the atlas is relatively small, which is why we are able, as in the case of the women in Fig. 1-24 on page 32, to carry such a large proportion of our body weight on our heads without compressing the vertebrae (Fig. 5-20).

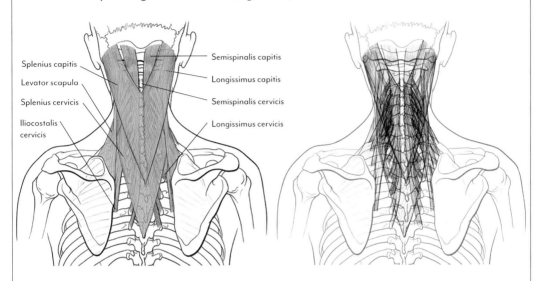

Splenius capitis

Levator scapula

Splenius cervicis

Iliocostalis cervicis

Semispinalis capitis

Longissimus capitis

Semispinalis cervicis

Longissimus cervicis

Fig. 5-19. The muscles of the cervical spine can be represented as prestressed cables in a complex tensegrity structure where they act on the bony skeletal parts.

Fig. 5-20. With the spine and connective tissues functioning as a tensegrity structure, the pressure of the head on the atlas of the spine is relatively low.

Consider also what happens when you are bending, as in this inclined "monkey" position (Fig. 5-21). With the legs bent and the trunk inclined at the hips, the spine, far from supporting weight, must itself be supported. Four-footed animals, too, are in a position in which the legs are flexed, with the horizontal spine acting as a bridge between fore and hind limbs. In this position, the weight of the trunk is carried largely by the tensile members of muscles and connective tissues, which distribute the forces required to maintain postural support. In both cases, the spine is clearly not a weight-supporting column but part of a complex tensegrity structure in which the weight of the body is distributed within the tensile members and not carried by bones. Seen from this vantage point, it becomes clear that the human spine is much more than a compression-resistant column. It is a constantly moving variation of a more primitive animal design—one that preserves the basic tensegrity principles found in other animals, but in a complex and refined form in which the spine also functions, in the upright posture, as a compression-resistant column with the head sitting on top and the trunk balanced vertically on two legs.

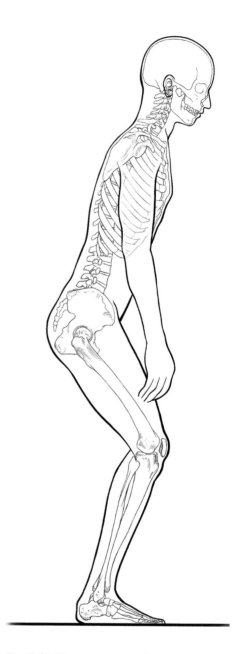

Fig. 5-21. The spine is more than a compression structure.

UPRIGHT POSTURE AND THE MULTIFIDUS MUSCLE

One of the key postural muscles of the spine is the multifidus, which fills up the grooves on either side of the spinous processes (Fig. 5-22). These deep postural muscles form an essential part of the network of muscles that maintain erect posture. They not only aid in the bending and rotating of the spine—in other words, in active movement—but also in maintaining the support of the vertebral column. As Gray writes,

> The multifidus spinae acts successively upon the different parts of the spine; thus, the sacrum furnishes a fixed point from which the fasciculi of this muscle act upon the lumbar region; these then become the fixed points for the fasciculi moving the dorsal region; and so on throughout the entire length of the spine; it is by the successive contraction and relaxation of the successive fasciculi of this muscle and other muscles that the spine preserves the erect posture without the fatigue that would necessarily have been produced had this position been maintained by the action of a single muscle.[2]

In this way, the action of the small muscles of the back, with the help and support of the ligaments, has the overall effect of straightening or elongating the spine.

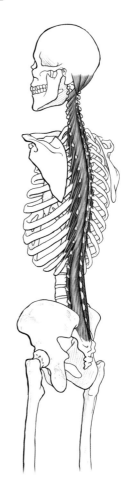

Fig. 5-22. The small postural muscles of the spine.

MOTOR UNITS AND ASYNCHRONOUS STIMULATION

When we sit or stand, muscles must work for extended periods of time—especially the deeper postural muscles, like multifidus. Even when these muscles are working, however, not all the muscle fibers that make up the muscle are active. Here's how it works.

A single skeletal muscle is made up of hundreds and sometimes thousands of muscle fibers. These fibers are served by many nerves, each of which serves a few dozen or, in the case of larger muscles, several hundred muscle fibers. The muscle fibers served by a single motor nerve are never concentrated in one part of a muscle but, for the sake of efficiency, are spread throughout the muscle. The group of muscle fibers served by a single nerve is called a *motor unit* because, when that motor nerve fires, it fires off all the muscle cells served by that nerve.

During forceful activity, all of the motor units within a muscle fire, activating all the muscle fibers within the muscle. In activities requiring background muscle tone, however, only some of the motor units—usually the deeper, slow-twitch fibers, which are designed to do less work over longer periods and are thus suited to maintain posture—are active, receiving a slow rate of impulses from the motor nerve in order to maintain background support. These motor units are also designed to work alternately so that, when the fibers in one motor unit become tired, another motor unit takes over, giving the first motor unit a rest—a phenomenon called *asynchronous stimulation.*

In Fig. 5-23, you can see how easily and effortlessly this child sits due to the action of these deep postural muscles. When these muscles are working properly, as they usually are in young children, it is possible to sit for long periods, without any sense of effort or fatigue. In this situation, you have little sense that different muscle fibers are firing at different times: at a purely unconscious level, some motor units alternately switch on and some switch off so that, over time, the workload is maintained and different muscle fibers are given a chance to periodically rest. This process is accomplished

Fig. 5-23. Child with perfectly functioning spine/postural system.

quite seamlessly so that, provided the system is working as it is designed to work, we have no sense that any muscular work is going on at all. Even when we are doing nothing and just sitting, muscles all over the body are receiving the slow, rhythmic discharge of impulses from motor nerves that maintain the low-level activity, or tone, necessary for stabilizing the body, maintaining muscle support, and keeping muscles toned and active.

EXERCISE:
The "Monkey" Position

Another useful way to bring about length in muscles is by standing in the so-called *monkey position*, which is identical to the vertical stance except that the trunk is inclined forward at the hips. As with the vertical stance, the knees are bent; but because the trunk is inclined forward, there is a greater opposition between the head, hips, and knees, which is, in this sense, more active and dynamic than the vertical stance, and more challenging. Performed properly, the monkey position creates an opposition between the head, hips, and knees that encourages lengthening and release in the back muscles, the thighs and legs, and the trunk as a whole. Releasing these tensions allows the legs to lengthen out of the back, which in turn allows the back to lengthen more fully and helps to reinstate the lengthened working of the musculoskeletal system as a whole.

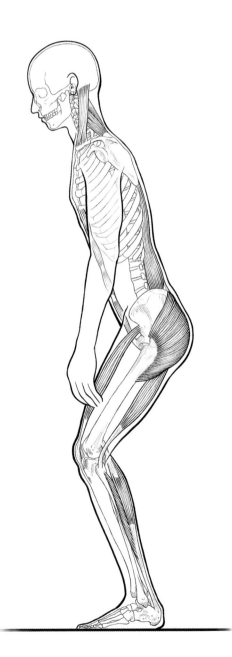

The monkey position (Fig. 5-24), as you will see when you try out the following procedures, places an increased demand on the extensor muscles of the back. When we are standing fully upright, the muscles of the back don't have to work very hard because all the body parts are stacked on top of one another. Because we habitually shorten and interfere with ourselves, many of the key supporting muscles in the body become too lax, forcing us to rely on

Fig. 5-24. The "monkey" position.

ligaments when we are slumping, and then to overcompensate by tightening the wrong muscles when we need to sit up. When we are in the monkey position with the trunk inclined forward, the back muscles have to work quite hard to keep the trunk from toppling forward. In this situation, one of two things can happen. The first is that, to maintain this position, we can tighten, shorten, and strain the back muscles, particularly in the lower back region. That is, of course, what most people do when bending. The second is that, by encouraging release and opening out, the back can become more elastic and open, so that the workload is distributed over the whole of the back, the supporting muscles become more toned and elastic, and we begin to restore the supporting function of the back muscles. This, of course, is what we want, and it is the entire point of practicing this procedure. By encouraging the lengthening action of the muscles of the neck, back, and legs, we produce a healthy, toned, elastic state of the musculature.

This elastic state of the muscles is a physical condition, but it is brought about as the result of a thought process called *directing*. When we are able to think or direct effectively in this way, we bring about a state of fluidity and release in the muscular system, which is the condition in which the muscles work most efficiently, and which is particularly useful in restoring the proper working of the extensor muscles of the back.

Before bringing about more length in muscles by "directing" body parts, how-ever, we have to think about how to go into the monkey position. Because we tend to collapse and shorten the body when we bend, it is easy to go into this position by pulling ourselves downward in front and losing length in the trunk. This defeats the purpose of the exercise, since we want to prevent shortening, to maintain lightness, and to achieve as much opposition of body parts as possible. As a first step, then, we must think carefully about how to go into monkey, taking care to stay light and to have an upward flow in the trunk, not to stiffen in the joints, and not to bend or fold into the monkey or to sink into the legs. If we succeed, the monkey is then light and free, and we have a much better chance of bringing about release once we are in the position.

Going into Monkey Position

When we go into the monkey position, there are two tendencies to avoid. The first is folding or bending in the hips; instead of deliberately folding ourselves into position, we must think of releasing to allow the body to incline forward at the hips. In this way, the bending happens as a result of releasing, not because we are pulling ourselves into position.

The second tendency we want to avoid is dropping or pulling down in front and sinking into our legs. We want instead to come to our full stature and then, when we initiate the movement, to think of letting our head lead as we fall upward to go into monkey. Again, we do not want to think of bending and instead want to maintain a sense of length while releasing in our joints as the basis for moving. Then go into the monkey position as follows:

1. Stand with your feet apart. Think of allowing your feet to open onto the ground so that you can come up to your full stature. You want to be balanced over your feet, slightly back on your heels so that you aren't gripping in the ankles and legs, and allow yourself to come to your full stature by getting as much length in the back and front as you can (Fig. 5-25a).

2. Free your neck to allow your head to tip forward, but in such a way that the head goes up. The movement of going into monkey is then initiated by this forward and up movement of the head.

3. Think of releasing the fronts of the ankles, the backs of the knees, and the front of the hip joints.

4. Continuing to think these directions, allow your knees to release from your lower back right around the buttocks and hamstrings to the backs of your knees so that your knees come well forward over your feet as you incline into monkey position (Fig. 5-25b). Don't forget, as you do this, to make sure the head leads in the other direction—otherwise you will sink into the legs. This should bring you into a balanced monkey, as shown in the illustration.

"Directing" in Monkey Position

Once you are able to go into this position in a balanced way, you want to stop—that is, not alter the position in any way—and give the following directions:

1. Think of lengthening the back from head to hips by allowing your head to go away from your hips and your hips to go away from your head (Fig. 5-25c).

2. Think of allowing your body to have its full length in front from the front of the hip bones (the anterior iliac crest of the hips) to the mastoid process just behind the ear (Fig. 5-25c).

3. Think of allowing your knees to release away from your hips by releasing in the ankle joints and asking for length along the thighs. Think also of releasing in your calf muscles to let the knees release forward from your heels, so that your

knees go forward and away from your hips and heels as you release the thighs and the calf muscles (Fig. 5-25c).

Again, the directions are thoughts that bring about release, not actual movements that we are trying to bring about deliberately. This means that you must not attempt to "do" the directions but only to "think" them, which brings about an internal release of the muscles of the neck, back, thighs, and calves.

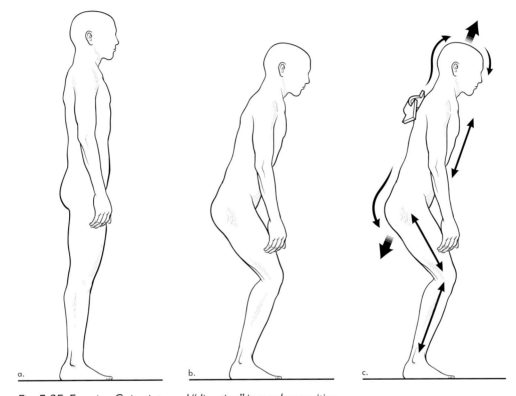

Fig. 5-25. Exercise: Going into and "directing" in monkey position.

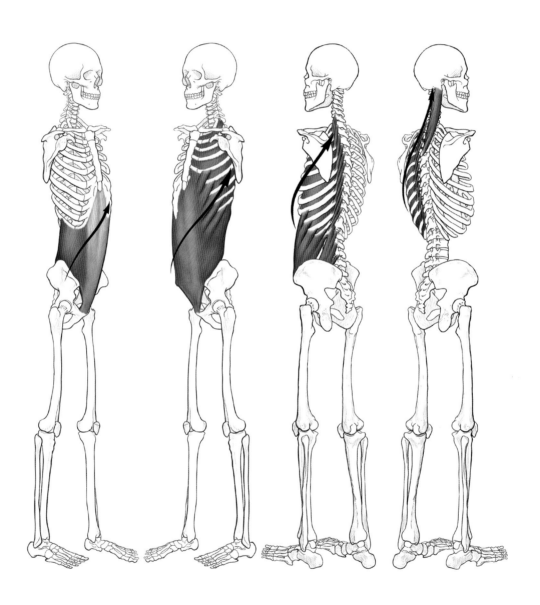

6

The Spiral Musculature of the Trunk

The trunk is encircled by *oblique muscles* that, originating at the pelvis, wrap around the abdomen and ribs and attach to the skull. These muscles facilitate torsional movements of the head and trunk, making it possible to efficiently rotate the head and trunk around the spinal axis as the foundation for complex movements.

Key to the Spiral Musculature

Although the oblique muscles of the trunk torque the body, they are part of a larger system of support in which the head and trunk, rotating around the spinal axis, maintain lengthened support against gravity. The key to this system, as with the extensors in the back and the flexors in front, is the forward balance of the head, which counterbalances the action of the neck and back muscles, in relation to the lengthening spine. When working properly, the spiral muscles do not simply wrap around the trunk but form a suspension system in which the trunk and spine hang from the head so that torsional movement can take place with a maximum of freedom and poise.

Contraindications

When we habitually twist or torque the body, the muscles of the neck, shoulders, trunk, and hips become chronically tense. The neck muscles are shortened and the head asymmetrically fixed in position, and the constant use of the dominant arm causes the flexors of the dominant limb to become shortened and overworked.

WHEN WE ARE IN A POSTURAL TWIST

1. The head is pulled back.
2. The intercostal muscles are shortened and the ribs are fixed.
3. There is a loss of front length.
4 The trunk as a whole is twisted and shortened from skull to pelvic rim.

WHEN THE SPIRALS ARE LENGTHENING

1. The neck muscles release.
2. The head is balanced forward and up.
3. The flexor sheets of the abdomen and thorax lengthen.
4. The spine and trunk lengthen and "untwist" from skull to pelvic rim.

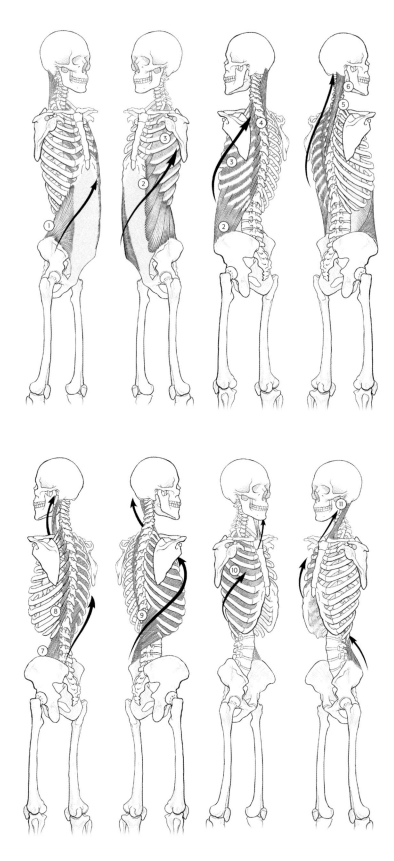

ORIGIN	MUSCLE	INSERTION
FRONT SPIRAL		
Iliac crest/lumbar fascia	**1. Internal abdominal oblique**	Pubic bone/costal arch, Ribs 1–12
Ribs 5–12	**2. External abdominal oblique**	Iliac crest
Lower border of rib above	**3. External intercostal**	Rib below
Transverse process	**4. Levatores costarum**	Two ribs below
Nuchal ligament/T3–T6	**5. Splenius cervicis**	C1–C3
C7–T3	**6. Splenius capitis**	Mastoid process/occiput
BACK SPIRAL		
Iliac crest	**7. Quadratus lumborum**	L1–L5, T1
Transverse process	**8. Levatores costarum**	Two ribs below
Spinous processes, T11–L3	**9. Serratus posterior inferior**	Ribs 9–12
Lower border of rib above	**10. Internal intercostal**	Rib below
Sternum/clavicle	**11. Sternocleidomastoid**	Mastoid process

The Spiral Lines

Although the simple flexor line we looked at in Chapter 4 runs vertically from the pubic bone to the skull, most of the flexors run in oblique lines that permit twisting movements of the trunk. These oblique muscles are confined to each side of the body and therefore do not cross the midline in front or back. When the direction of pull is followed across the midline, however, these muscles, transmitting forces across fascial sheets, can be seen to form continuous spirals wrapping around the trunk from head to pelvis. First described by the anatomist Raymond Dart (see the sidebar titled "Raymond Dart and the Double Spiral Design" on page 116), these spirals form a double helix suspensory arrangement in which the trunk is suspended from the head (Fig. 6-1).

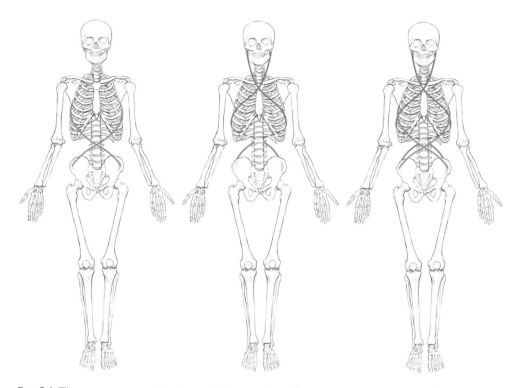

Fig. 6-1. The arrangement of the four spirals, two in front (in green—one on either side) and two in back (blue—one on either side), reveals a complete double helix.

Tracing the Spiral Lines: How the Oblique Layers Form a Double-Helix Spiral Pattern

When viewed in discrete pieces, the oblique muscles seem to form a random patchwork of muscles. But when you piece together a continuous pattern by identifying an oblique muscle running in a particular direction in one layer and then another muscle layer on the other side of the midline that continues in the same direction, it is possible to trace a continuous spiral that begins at the pelvis, wraps around the trunk, and ends up at the base of the skull (Fig. 6-2).

This can be done in both directions, so that we can trace two complete spirals running from the pelvis right up to the head, one on each side, forming a complex double-spiral that encircles the body and continues even into the limbs, making complex rotational movements possible.

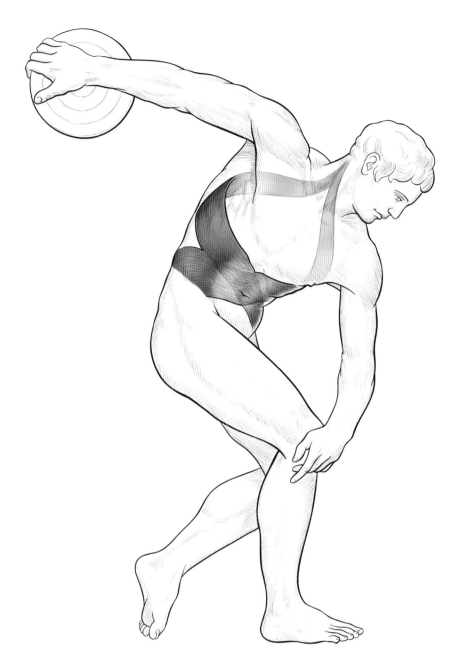

Fig. 6-2. On this discus thrower, we clearly observe the overlapping of the spiral musculature in front (highlighted) from the pelvis to the skull.

RAYMOND DART AND THE DOUBLE SPIRAL DESIGN

The double-helix design of the human musculoskeletal system was first described by Raymond Dart, the world-renowned anthropologist who is credited with having discovered the first early hominid skull. Dart began his career as a doctor and anatomist and, as a young man, took a post as Professor of Anatomy at the University of Witwatersrand in Johannesburg, South Africa. While there, he pursued a side interest in anthropology and evolution and asked for skeleton fossils, which were dug up from time to time at a local quarry, to be brought to him for examination. One day, when observing a group of fossils that had just been delivered to his home, he recognized a small skull that appeared to belong to a chimpanzee or baboon, but which had human-like teeth. He painstakingly chipped away the rock and, on further examination, was convinced that the skull belonged not to a baboon but an early human. When he published his findings, his claim that this small-skulled creature was an early hominid species—and that this species had evolved not in Europe but in Africa—drew skepticism and scorn. When further skulls were discovered, he was later vindicated and became famous for his groundbreaking discovery.

Dart's description of the double-helix muscular arrangement of the human muscular system grew out of a related but more urgent interest. His daughter suffered from night terrors and his infant son, who had been born premature, was partially spastic. In searching for help for his children—and to address a scoliosis that had given him trouble for a number of years—he had lessons in the Alexander Technique. To further his study, he began a personal exploration of human anatomical design, wrote several papers on the Alexander Technique and, over the next few years, developed a series of developmental movements designed to improve bodily poise and coordination, now known as the *Dart Procedures*.

Perhaps Dart's most interesting and innovative observation was that the oblique muscles of the human body were in fact part of a larger spiral design. In a paper titled "Voluntary Musculature in the Human Body: The Double Spiral Arrangement," Dart observed that the oblique muscles of the abdomen and thorax, when pieced together, formed a double-spiral that could be traced from the pelvic rim to the skull—a global feature of our upright design that was missed because, he said, anatomists focused on the details and overlooked the larger functional design (Fig. 6-3a and b). Rotating beneath the skull, the trunk could be said to be suspended, via fascial tissue, by these oblique muscle sheets. "Thus, in a very real sense, the occiput and spines of the vertebrae suspend the body by means of two spiral sheets of muscle encircling the trunk."[1] It was this spiral design, Dart argued, that made it possible to perform complex movements, but it could also contribute to postural twisting, which could be addressed by placing oneself in positions that gave the postural twist a chance to unwind and thus restore bodily poise.

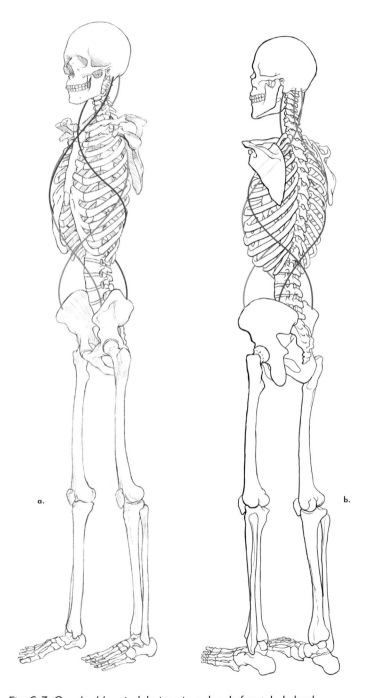

Fig. 6-3. Our double spiral design viewed: a. In front; b. In back.

The Front Spiral

The internal abdominal oblique muscle originates at the anterior rim of the pelvis and runs obliquely upward to the midline (Fig. 6-4a). If you cross the midline and continue in the same direction, you are following the external abdominal oblique muscle, which continues into the rib cage as the external intercostal muscle (Fig. 6-4b and c). This muscle wraps around the rib cage and intersects with the levatores costarum and the transverse processes of the cervical vertebrae at the midline in back (Fig. 6-4d). If

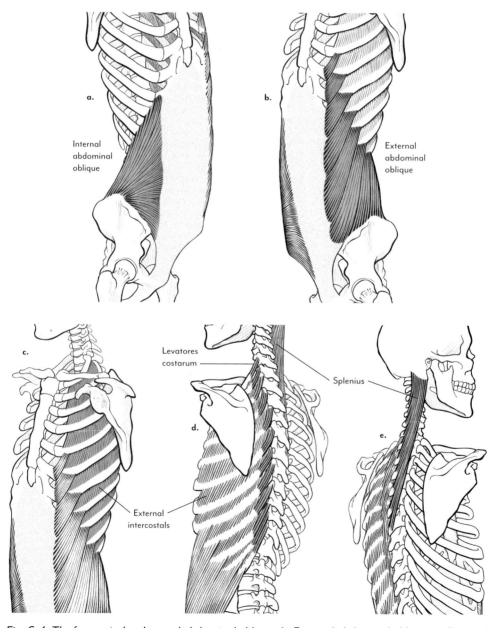

Fig. 6-4. The front spiral: a. Internal abdominal oblique; b. External abdominal oblique; c. External intercostals; d. Levatores costarum; e. Splenius.

we cross the midline of the spine, we can continue in this direction with the splenius muscle, which originates at the transverse processes on the right side of the upper spine and runs obliquely upward, attaching to the right occiput of the skull—the same side on which the spiral began (Fig. 6-4e). We thus have a spiral line that begins at the front rim of the pelvis and wraps around the body to end up at the back of the skull.

So, beginning from the pelvis, it is possible to trace a continuous spiral that begins at the right anterior rim of the pelvis, crosses the abdomen to the rib cage on the left side, circles around the ribs to the back, continues obliquely across the transverse processes of the cervical vertebrae in back, and finally ends at the occiput on the same side as it began.

You can trace the same spiral starting on the left side, so that you have two spirals in front—one on the right and one on the left—forming a kind of double spiral arrangement of muscles wrapping right around the torso, starting from the pelvic rim and ending at the skull.

The Back Spiral

You can also trace two spirals that start at either side of the pelvis in back. Each one of these begins at the pelvis rim, following quadratus lumborum to the levatores costarum to the spine (Fig. 6-5a). On the other side of the spine, the line continues with the serratus posterior inferior (Fig. 6-5b) as well as the internal intercostal muscle (Fig. 6-5c and d), which wraps around the rib cage to the sternum and, crossing the midline, follows the sternocleidomastoid muscle to the mastoid process of the skull (Fig. 6-5e). This makes a total of four spirals, two in back starting on either side, and two in front (see Fig. 6-3).

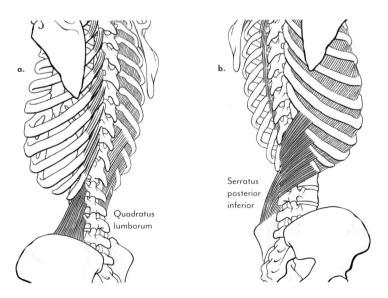

Fig. 6-5. The back spiral: a. Quadratus lumborum; b. Serratus posterior inferior.

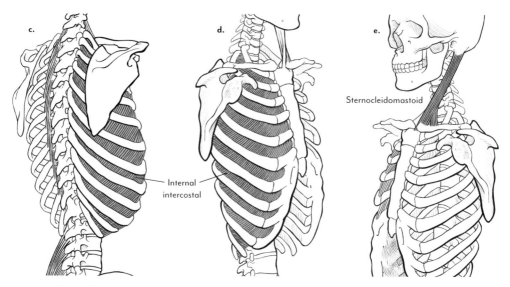

Fig. 6-5. (cont.): The back spiral: c. and d. Internal intercostals; e. Sternocleidomastoid muscle.

The Human Architectural Marvel

No other animal's structure compares to the moving architectural structure of the upright human design. A cat is capable of complex twisting movements, as we can see when it is dropped back down and gyrates its body to land on its feet (Fig. 6-6). No vertebrate, however, is capable of twisting movements as complex as those of humans. The oblique muscles that wrap around the trunk form a double spiral arrangement that makes it possible to rotate the head and body around a vertical axis—an amazing development that has granted human beings a 360-degree field of vision, the ability to walk with a striding gait, and the possibility of mastering a huge array of skilled activities (Fig. 6-7). In our upright human posture, the musculature is organized around a vertical axis with the head sitting on top.

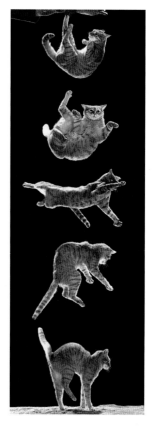

Fig. 6-6. Twisting movements of a falling cat.

We need only to watch a dancer executing a perfect pirouette, a golfer swinging a club, or a martial artist gyrating in the air to land a well-placed kick on an opponent to understand that, in movement, the body can perform complex twisting movements with the head sitting on top and the body slung from the head (Fig. 6-8). We are not merely stacked up or aligned over our feet; we are also, in a sense, suspended from the skull, with the musculature of the trunk and the legs arranged underneath in a double-spiral—an architectural marvel unequalled in all of nature.

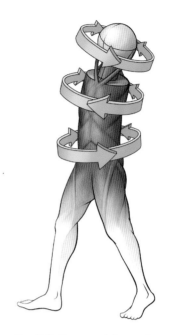

Fig. 6-7. Our spiral design, from trunk to head, enables complex twisting movements around a vertical axis.

Fig. 6-8. This Flamenco dancer twists her torso to the left and her pelvis to her right as she twirls.

THE SPIRALS WRAPPING AROUND THE CYLINDER

A simple way of understanding the spirals is to think of the trunk as a kind of cylinder, with the iliac crests of the pelvis forming the bottom rim of the cylinder and the base of the skull forming the top. Starting from the pelvic rim, each spiral begins on one side of the bottom rim of the cylinder and, wrapping around the trunk, attaches to the skull, or upper end of the cylinder, on the same side that it began.

Given this body plan, we can see why the most important spiral muscles are flexors and not extensors. In Chapter 4, we saw that, because the gut and rib cage lie in front of the spine, the flexors do not, like the extensors, lie close to the spine but instead wrap all the way around the abdomen and rib cage. Given our roughly cylindrical shape, the muscles that would be most effective in producing rotation are the ones that wrap around it—in other words, the flexors that encircle the abdomen and rib cage (Fig. 6-9).

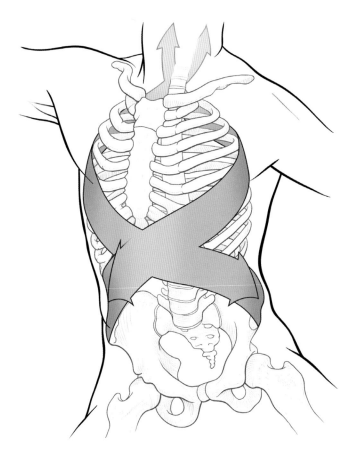

Fig. 6-9. The flexor muscles, shown here, contribute more to the spirals than the extensor muscles.

POSTURAL TWISTS

Although we are designed to perform twisting movements, constant rotation or torquing can cause the body to become habitually twisted. We can observe such postural twisting in a child learning to write or a programmer working in front of a computer; if you are right-handed, the usual tendency is to bring the right shoulder forward and twist to the left (Fig. 6-10). Some degree of spinal twisting is present in virtually all of us, particularly if we are engaged in manual skills that involve repetitive use of the dominant hand and asymmetrical twisting of the body. When such twisting is chronic and actions become strained, this causes undue activity in the flexors, which can lead to various problems with the shoulder, arm, wrist, and hand (and with the hips and legs). Many specific musculoskeletal complaints can be traced to postural twists, which can be treated only by untwisting the body so that the activity of the flexors is reduced and the arm and shoulder can again function normally.

Fig. 6-10. Postural twist in a child writing—note that
he faces sideways and that his right shoulder is raised.

Although massage and bodywork can provide relief from postural twisting, the only way to restore normal function is to release the chronically shortened muscles that are causing the twist, allowing the body to untwist and to regain its full lengthened support. Since the spiral muscles that wrap around the trunk ultimately attach to the head, the key to undoing postural twists is to encourage the lengthened support of the head and trunk, which then allows the untwisting to take place. Even postural twists that seem to center on the hip or shoulder always involve shortening and twisting throughout the entire muscular system, which is why the key to untwisting is to restore dynamic length to the entire system, with the head leading and the back and trunk lengthening.

THE EVOLUTION OF THE DOUBLE-SPIRAL MUSCLES

Although it is fairly easy to trace the spiral muscles that wrap around the trunk, it is not as easy to understand how—and why—this design came about. The simple answer is that the double-spiral musculature in humans originates with the first vertebrates that came onto land. In a fish, muscles are arranged segmentally along either side of the body to produce lateral flexion in the water. Animals such as worms can combine action of different segments to produce crude twisting movement. Vertebrates that came onto land needed to move in more sophisticated ways. To make rotational movements possible, the muscles of the trunk divided into three layers. The deepest layer, the transverse abdominal muscle, runs horizontally; the middle and superficial layers run in oblique directions. The splitting of the abdominal and thoracic muscles into layers now produced two spirals, running in opposite directions, that wrapped around the body, making it possible to twist the body very efficiently around the vertical axis.

We can see these stages clearly unfold in the developing embryo. In the earliest stages of embryonic development, the proliferating cells form into an elongated structure and begin to differentiate into three germ layers: the *endoderm* (the gut), the *ectoderm* (nervous system), and the *mesoderm* (the musculoskeletal system and related organs) (Fig. 6-11). In vertebrates, the mesoderm, or muscle tissue, begins to organize in segments on each side of the primitive spine, or *notochord*. Clefts appear on each side of the notochord and the mesoderm tissue breaks into distinct clumps, or *somites*. These paired blocks of tissue—called *paraxial mesoderm* because they are arranged alongside the notochord—contribute to skin, muscles, cartilage, and bone to produce the basic vertebrate body plan (Fig. 6-12).

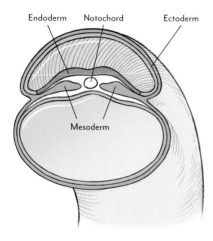

Fig. 6-11. Embryonic germ layers.

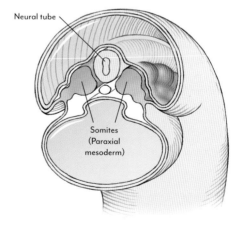

Fig. 6-12. Development of somites in the embryo.

At this point in its development, the embryo is more or less at the fish stage of vertebrate evolution. There is a head at the front end, a central notochord, or primitive spine, forming the long axis of the body, with muscles arranged in segments on either side and, corresponding to these segments, the peripheral nerves will innervate each segment to produce lateral flexion in the water (Fig. 6-13).

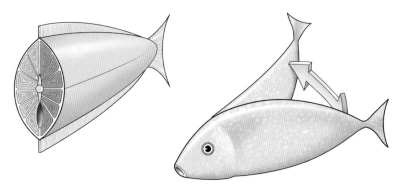

Fig. 6-13. Alternate contraction of muscle segments on either side of the spine result in lateral flexion in the fish.

As the embryo develops further, the somites divide into flexor and extensor halves, served by extensor and flexor divisions in the peripheral nerves that innervate each system. The embryo is now developing limbs and clearly corresponds to a terrestrial vertebrate. It has also developed muscle divisions capable not only of lateral flexion but of flexing and extending the body forward and back (Fig. 6-14). The body plan is now characterized by two fundamental divisions that we see in all land vertebrates: the bilateral division into left and right halves based on segmented mesodermal tissue that will become muscles, vertebrae, ribs, and skin, served by peripheral nerves feeding each segment; and the division of the muscles into dorsal and ventral halves, served by corresponding flexor and extensor nerve branches. Even the complex human design conforms to this basic body plan: we have muscles on the right side and muscles on the left and never the twain shall cross. Even the mouth, which appears to cross the midline, is in fact made up of a right and left half; and the diaphragm, which also bisects the midline, is composed of tissues on both sides. The

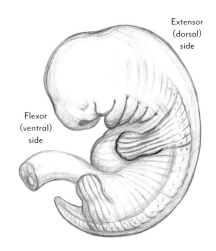

Extensor (dorsal) side

Flexor (ventral) side

Fig. 6-14. Embryo with flexor/extensor division.

same is true of the extensor/flexor division: muscles are one or the other and do not cross this dividing line.

To produce rotational movements, however, these land vertebrates needed muscles that could transmit forces across these dividing lines, and at oblique angles that could produce twisting movements. We see how the problem is solved in the developing embryo. Immediately after the flexor division appears, flexors on each side begin to stratify into layers that form interlacing sheets running in oblique directions. The deepest layer, the transverse abdominal muscle, runs horizontally; the middle layer, or internal abdominal oblique, slants in one direction; and the superficial layer, or external abdominal oblique, slants in the other direction. These layered muscles appear within existing segments—flexors confined to the right half and flexors confined to the left—but they produce spiral muscles on both sides of the body that can work together as functional groups, making it possible to twist the body efficiently around the spinal axis (Fig. 6-15).

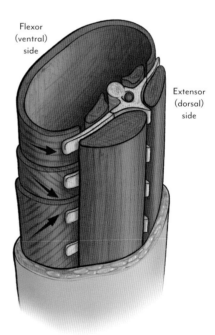

Flexor
(ventral)
side

Extensor
(dorsal)
side

*Fig. 6-15. Stratification of the flexors
into oblique layers.*

HOW THE TRUNK IS SUSPENDED FROM THE HEAD: OUR AMAZING DOUBLE-SPIRAL DESIGN

It is easy to understand that muscles spiral around the trunk, attaching ultimately to the head, but what does it mean to say that the body is suspended from the head by means of these spirals? The first vertebrate—the fish—moves toward food by levering its spine laterally with the aid of muscles arranged along each side of the body. Animals that came onto land needed to support themselves up off the ground in order to use their limbs for locomotion. The spine then becomes a horizontal bridge with the head extending out from the body. Extensor muscles on the neck support the head while extensors on the back support the body on extended limbs. Amphibians and reptiles flex the body sideways to advance their limbs; mammals are able to coil and extend their bodies to push off with their legs. But the relationship of the head to the body organizes movement regardless, and even in mammals, the body essentially swings from the head, as we can see when a dog wags its tail and the tail wags the body, or when it runs (Fig. 6-16).

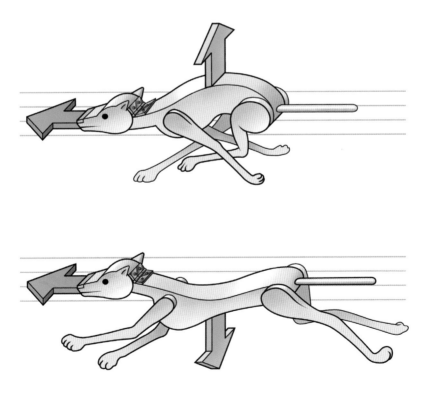

Fig. 6-16. In a four-footed animal, the body "swings" up and down from the head, while the head remains in the same plane as the direction of movement.

In humans, this arrangement is altered yet again. The spine and trunk are up-ended and the head, which no longer hangs out in front of the spine but now sits on top of it, is delicately poised off-center on the spine to counter the backward pull of the extensor muscles. The body still swings from the skull, just as in a fish or four-footed animal, and the relationship of the head to the trunk remains the primary factor organizing the upright support system.

One other crucial change occurred in the process of coming to the upright posture. Because the trunk is now vertically poised on two feet with the head balanced on top of the spine, the oblique muscles wrapping around the trunk made it possible to rotate the body around a vertical axis—an all-important development that granted human beings a 360-degree field of vision, the ability to walk with a striding gait, and the possibility of mastering a huge array of skilled activities. The structure is supported from the ground up because we are stacked over our feet and must apply downward force to come up from the ground. But the body must also be organized from the top down, because it cannot maintain lengthened and efficient support against gravity unless the head is counterbalanced on top of the spine, with the muscles of the neck and trunk lengthening to allow the head to go upward. In our upright human posture, we are not merely stacked up and aligned over our feet; the trunk is actually suspended from the head in a double-spiral arrangement—an altogether extraordinary design (Fig. 6-17).

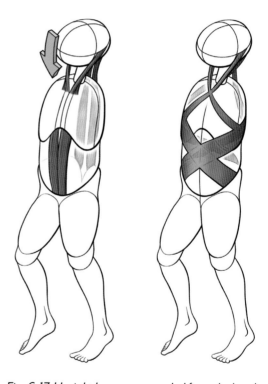

Fig. 6-17. Upright human suspended from the head.

EXERCISE:
How to "Undo" a Postural Twist

One of the simplest things you can do to come out of a postural twist is to lie down in the semi-supine position with some books under your head and your knees up. We have already seen on pages 59 and 83 how you can "direct" your head, trunk, and knees. In this position, the head and trunk are fully supported, which gives chronically contracted muscles of the neck and trunk a chance to release so that the body can untwist into its fully lengthened state (Fig. 6-18).

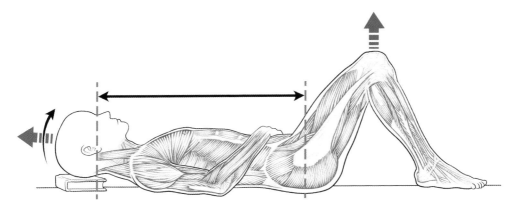

Fig. 6-18. Semi-supine position: undoing a postural twist.

When chronically tight muscles let go, this can bring considerable relief and therefore can feel like a treatment or cure. But coming out of a postural twist produces a state of lengthened support that applies directly to balance and movement on two feet and is therefore much more than a cure or form of relief. This is why one cannot address a postural twist simply by trying to release or treat muscles. Twisting is something we do in activity and something that we have to stop doing, which takes time, patience, and skill. As muscles let go and the head goes out of the back, one can feel an untwisting in the neck, ribs, and trunk as the body regains its state of natural poise.

When the spiral muscles release out of a chronic twist, two key areas need to let go. The first is the pectoral/biceps region of the shoulder. Because postural twists are usually triggered by the use of our dominant hand, the biceps and pectoral muscles on the dominant side typically become shortened; the release of these muscles brings about a widening of the shoulders and reduction of tension in the flexors of the arms, sometimes accompanied by a kind of toning and release in the hands and fingers. The second area that releases is the hips, which often become clenched in the buttocks region, particularly on the dominant side. When this area lets go, the outward pull of the gluteal muscles is reduced, the hips open up, and the right leg, which is often pulled or rotated outward, rotates inward. This area, which we look at in more detail in Chapters 11 and 12, is directly related to the spiral musculature of the legs, which reflect the spiral musculature of the trunk.

Although specific areas release when the body untwists, it is important to mention that these changes cannot be brought about directly but are the indirect result of restoring length to the system as a whole. It is also counterproductive to try to counter a postural twist by rotating the other way, which only complicates matters further. One must learn how the body releases into length so that shortened muscles in specific areas (such as pectoral/biceps region) can let go. This untwisting can sometimes feel uncomfortable and even wrong—when one is habitually twisted to the left, the untwisting may feel as if you are now twisting to the right!—but if you trust the process, the result is a dramatic improvement in function and poise.

For more information on directing, see the appendix on page 287.

EXERCISE:
Tracing the Spirals

To trace the spirals, partner up with a fellow student or colleague; your partner will be the student and, as the teacher, you will trace the spiral lines of the trunk, beginning at your student's pelvis and ending at their skull. With your student lying down in the semi-supine position, stand where indicated by the X.

The Front Spiral (left)

Your student lies on the table in semi-supine position and you stand at your student's right side (indicated by the "X" in Fig. 6-19).[2]

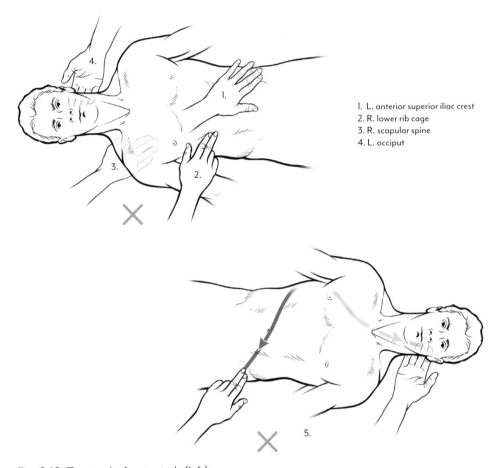

1. L. anterior superior iliac crest
2. R. lower rib cage
3. R. scapular spine
4. L. occiput

Fig. 6-19. Tracing the front spirals (left).

1. Place your left hand on your student's left hip bone (anterior superior iliac crest).

2. Place your right hand on your student's right lower rib cage. The student thinks of lengthening between these two points.

3. Move your left hand to the right side of your student's scapula in back near the neck.

4. Move your right hand to the left side of your student's occiput.

5. Move to the left side of your student and place your left hand on your student's left hip bone and your right hand on the occiput.

The Front Spiral (right)

The student lies on the table in semi-supine position and you stand at their left side (Fig. 6-20).

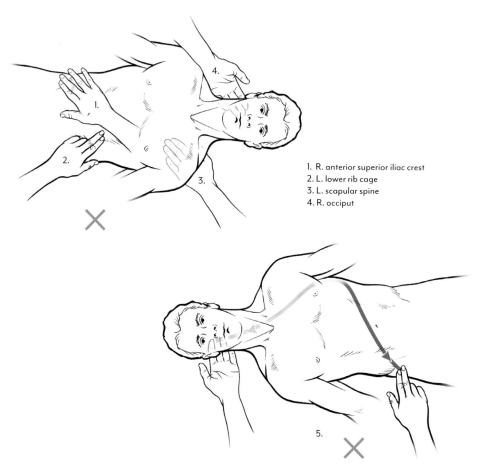

1. R. anterior superior iliac crest
2. L. lower rib cage
3. L. scapular spine
4. R. occiput

Fig. 6-20. Tracing the front spirals (right).

1. Place your right hand on your student's right hip bone (anterior superior iliac crest).

2. Place your left hand at your student's left lower rib cage. The student thinks of lengthening between these two points.

3. Move your right hand to the right side of your student's scapula in the back, near the neck.

4. Move your left hand to the student's right occiput.

5. Move to the right side of your student and place your right hand on your student's right hip bone and your left hand on the right occiput.

The Back Spiral (left)

The student lies on the table in the semi-supine position and you stand at their right side (Fig. 6-21).

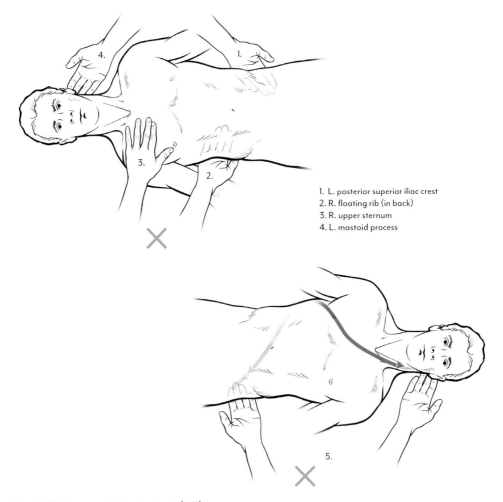

1. L. posterior superior iliac crest
2. R. floating rib (in back)
3. R. upper sternum
4. L. mastoid process

Fig. 6-21. Tracing the back spirals (left).

1. Place your left hand on the left back iliac crest (posterior superior iliac crest).

2. Place your right hand on the right floating rib in back.

3. Place your left hand on the right side of the sternum.

4. Place your right hand on the left mastoid process.

5. Go to your student's left side and place your left hand on the back of the left iliac crest and your right hand on the left mastoid process.

The Back Spiral (right)

The student lies on the table in semi-supine position and you stand on your student's left side (Fig. 6-22).

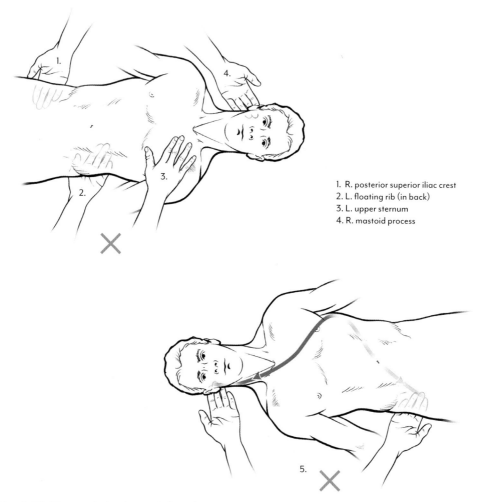

1. R. posterior superior iliac crest
2. L. floating rib (in back)
3. L. upper sternum
4. R. mastoid process

Fig. 6-22. Tracing the back spirals (right).

1. Place your right hand on the right back iliac crest (posterior superior iliac crest).
2. Place your left hand on the left floating rib in back.
3. Place your right hand on the left side of the sternum.
4. Place your left hand on the right mastoid process.
5. Go to your student's right side and place your right hand on the back of the right iliac crest and your left hand on the right mastoid process.

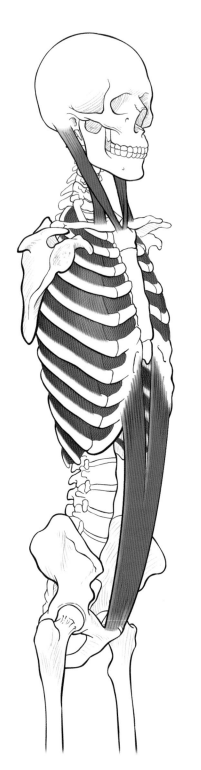
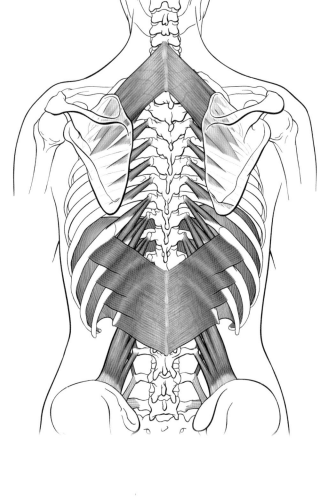

7

Breathing

When we breathe, air is taken into the lungs, oxygen from the air is absorbed into the blood stream, and the blood stream supplies cells throughout the body with oxygen. Although much of this process takes place internally, the flow of air into and out of the lungs is caused by movements of the ribs and diaphragm and is therefore dependent on the efficient working of the musculoskeletal system.

The Key to Breathing

Although we can control specifics aspects of our breathing, the ability to breathe efficiently depends on the flexibility and support of the postural system as a whole. When the trunk lengthens against gravity and the shoulders and back are full and supported, the ribs can move freely and the thorax has its full capacity.

Contraindications

Most of us interfere with our breathing by lifting the chest, pulling back our head, and shortening in stature in preparation to speak. When we vocalize, we then compress the ribs and chest, further interfering with the flexible support of the trunk and ribs. Because we are unconsciously taking and holding our breath, the diaphragm becomes chronically contracted, the ribs do not move flexibly, and breathing becomes generally impaired. These tendencies are compounded by heightened stress and high levels of tension in the body.

WHEN WE INTERFERE WITH BREATHING

1. The chest is collapsed and raised.
2. The ribs are tight.
3. The shoulders are narrowed.
4. The back is shortened and narrowed.
5. The diaphragm is chronically contracted and the ribs are raised.

WHEN THE BREATHING IS FREED UP

1. The chest and rib cage open up.
2. The shoulders are widened.
3. The back widens, the ribs move freely, and the costal joints regain mobility.
4. Full intrathoracic capacity is restored.
5. The diaphragm releases and ascends.

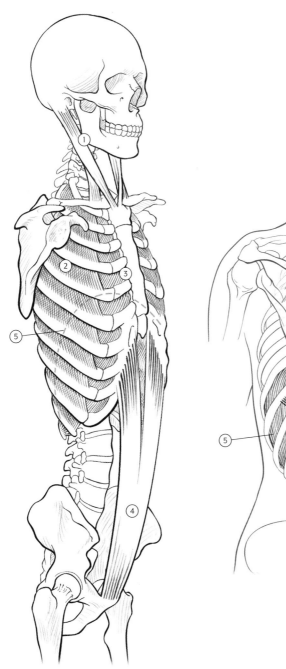
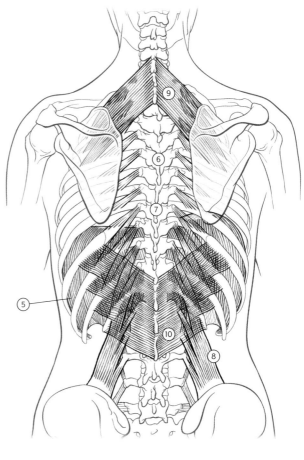

ORIGIN	MUSCLE	INSERTION
Sternum/clavicle	1. Sternocleidomastoid	Mastoid process
Lower border of rib	2. External intercostal	Upper border of adjacent rib
Lower border of rib	3. Internal intercostal	Upper border of adjacent rib
Lower sternum	4. Rectus abdominis	Pubic bone
Crus of lumbar vertebrae/sternum/costal arch	5. Diaphragm	Central tendon of dome of diaphragm
Transverse process	6. Levator costae	Rib below
Transverse process	7. Levatores costarum	Two ribs below
Iliac crest	8. Quadratus lumborum	L1–5, T1
C6, C7, T1, T2	9. Serratus posterior superior	Ribs 2–5
L1, L2, T11, T12	10. Serratus posterior inferior	Ribs 9–12

How We Breathe: The Ribs and Diaphragm

Breathing is one of the most vital of our life processes. All day long, throughout our lives, we take in air in order to provide cells throughout the body with oxygen, and then we expel carbon dioxide from the lungs in order to rid the body of wastes produced by cellular activity. Breathing—the flow of air into and out of our bodies—takes place by altering the size of our chest cavity. When the space within the chest cavity gets larger and smaller, air flows into and out of the chest through either the nose or the mouth. The efficiency and flexibility of these movements depend, in turn, on the upright support system.

We increase and decrease the size of the chest cavity in two ways. First, the ribs, which form the chest cavity, rise like pail handles by moving at the joints where they attach to the spine; this action increases the space within the chest (Fig. 7-1). The uppermost ribs connect in front to the sternum; those below form an arch beneath it; the last two, the floating ribs, do not attach in front. Because of this, not all the ribs move in the same way or to the same degree. But most of the ribs rise and widen to some extent, making the space inside the chest larger; when they return to their lower position, the space gets smaller.

Fig. 7-1. The action of the ribs in breathing can be likened to the movement of a pail handle.

Second, the bottom of the thoracic cavity is separated from the abdominal contents below by the dome-like muscle of the diaphragm which, by contracting, can flatten out and thus increase the size of the lower part of the chest cavity (Fig. 7-2). When the ribs rise and open, the diaphragm contracts and descends; the chest cavity increases and air rushes in to fill the lungs. When the ribs return to normal position, the diaphragm relaxes and ascends, air is forced out, and we exhale.

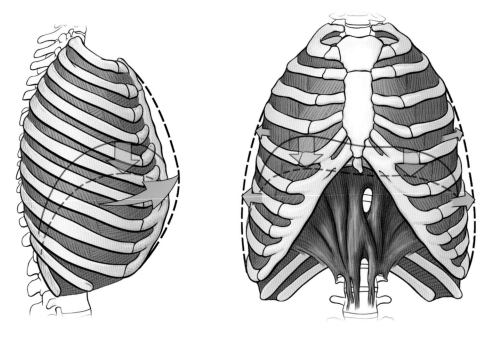

Fig. 7-2. The action of the ribs (dark blue arrows) and the diaphragm (light blue arrows) in breathing. As the ribs move upward and outward, the diaphragm moves downward.

The Anatomy of Breathing

The spine and rib cage form the basic framework for the respiratory system. As we've seen, the spine is made up of twenty-four moveable vertebrae—five in the lumbar region, twelve in the thoracic, and seven in the cervical. Twelve ribs on each side of the thoracic vertebrae form the rib cage (Fig. 7-3). The first seven ribs attach in front to the sternum, or breastbone; these are called *true ribs*. The remaining five are called *false ribs* because they do not attach directly to the sternum but join each other to form the *costal arch.* The final two do not attach in front but float. The ribs attaching to the sternum and the costal arch are not bony all the way around. At their extremity, the bone becomes cartilage, so where the ribs connect with the sternum and the costal arch is cartilaginous and quite flexible. The costal arch is also made up of cartilage.

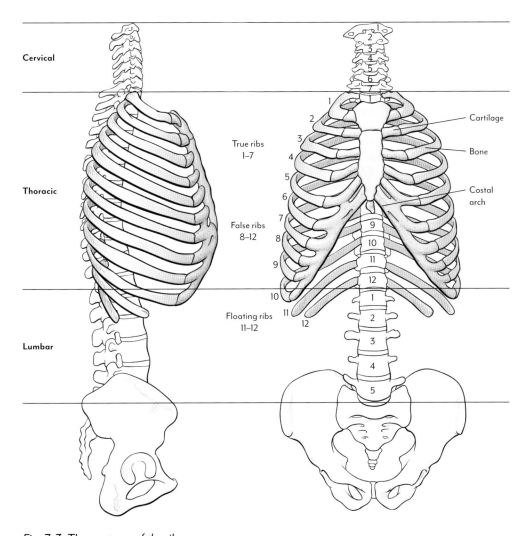

Fig. 7-3. The anatomy of the ribs.

The Joints of the Ribs

The articulation of the ribs with the spine permits the movements essential to breathing. Fig. 7-4 shows how each rib articulates with: a. the lower part of the body of one vertebra, b. the upper part of the one below it, and c. the disc in between the two vertebrae. The neck of the rib at (d) articulates with the transverse process of the lower of these two vertebrae. The rib is firmly bound at each of these articulations by several ligaments, permitting a limited rotation at the joint that nevertheless translates into quite a lot of movement over the entire length of the rib.

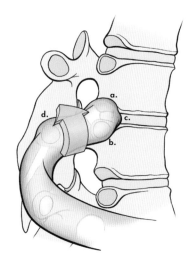

Fig. 7-4. The joints and movement of the ribs.

The Action of the Ribs

The movement of the ribs is crucial to breathing. Because the ribs slant down at an oblique angle, they hang below the point where they articulate with the spine. When we inhale, the ribs, by rotating where they articulate with the spine, move like pail handles being raised slightly. This rotation raises the sides of the ribs, which increases the lateral dimension of the thorax. It also brings the front of the rib forward as it moves upward, which increases the anteroposterior dimensions of the thorax as well. These movements increase the space within the thorax, causing air to flow into the lungs. Of course, not all the ribs move in the same way: the first ribs move very little, and there is in general more movement as you go lower down. The final two floating ribs, which are not attached either directly or indirectly to the sternum, have even greater mobility (Fig. 7-5).

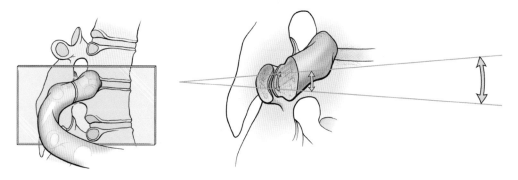

Fig. 7-5. Cross-section (green plane) of Fig. 7-4 at d. (the rib articulation with the transverse process), with arrows showing the rotation and upward movement of a rib.

Intercostal Muscles

Two sets of rib muscles, the internal and external intercostals, are directly responsible for movements of the ribs in breathing. There are eleven external intercostal muscles, one between each of the twelve ribs. They arise from the lower border of each rib and attach to the upper border of the rib below, running obliquely down and forward. Underneath this layer are the eleven internal intercostal muscles that arise from the inner surface of each rib and slant down and back, in the opposite direction to the external intercostals, to attach to the rib below (Fig. 7-6). These intercostals are the same muscles that we looked at in the previous chapter in relation to the body's spiral musculature.

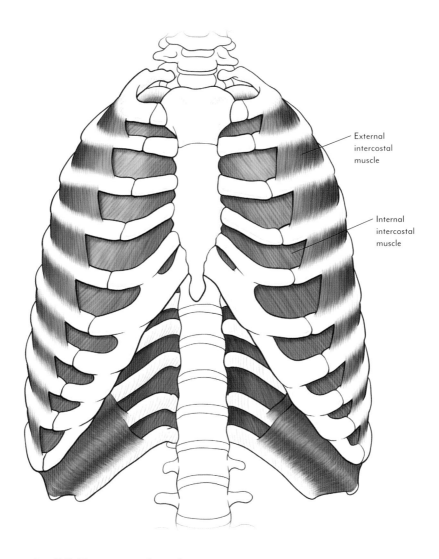

External
intercostal
muscle

Internal
intercostal
muscle

Fig. 7-6. The intercostal muscles.

The external intercostals function mainly to elevate the ribs, increasing the width of the thoracic cavity and causing inspiration. You can see from the angle of the muscle what the effect of contraction will be. When the upper ribs are held or fixed in place by the scalene muscles above, the contracting fibers pull from above to raise the ribs below, increasing the overall capacity of the chest. The internal intercostals act in the opposite direction, depressing the ribs when they contract, actively facilitating breathing out.

The levatores costarum and levator costae muscles originate at the transverse processes of the vertebrae and, running obliquely downward, attach to the ribs below (Fig. 7-7).

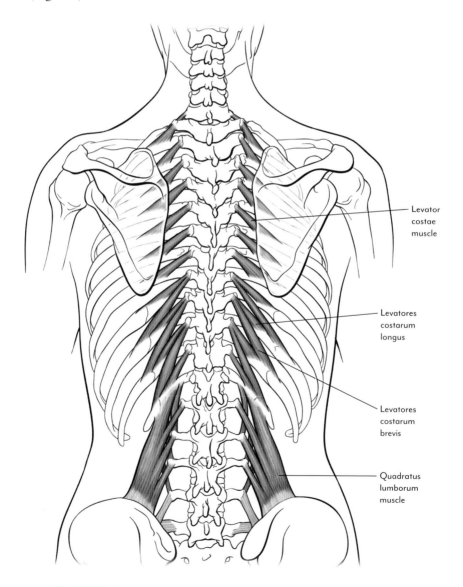

Levator costae muscle

Levatores costarum longus

Levatores costarum brevis

Quadratus lumborum muscle

Fig. 7-7. The levatores muscles in back.

These muscles, as their name suggests, assist the external intercostals in raising the ribs. When the quadratus lumborum muscle and levatores muscles are functioning properly, the lower back becomes elastic and filled out, and the floating ribs move freely. This coordinated lengthening and widening in the back tends to lend greater mobility to the ribs, which can then expand and move more freely.

The Diaphragm

The diaphragm is a large, dome-shaped muscle—actually, two domes. Its muscular fibers arise from the anterior spine and the entire circumference of the lower thorax and converge upward into a central tendinous peak that forms two domes on either side of this central tendon. The diaphragm's largest origin is a tendon, called the *crura*, which originates from both sides of the lumbar spine, the lower part of the sternum, from the costal arch formed by the lowest six or seven ribs, and from a ligament that spans the lower back from the first or second lumbar vertebrae to the lowest floating rib. From these points the muscular fibers of the diaphragm ascend in an arch to form a central tendinous aponeurosis at its top (Fig. 7-8).

The diaphragm contracts rhythmically every few seconds during normal breathing to ensure a constant supply of air to the lungs, the contractile portion of the diaphragm shortening and flattening out the central tendon and drawing it downward. At the same time, the ribs ascend, which also contributes to the expansion of the space within the chest cavity and the flow of air into the lungs.

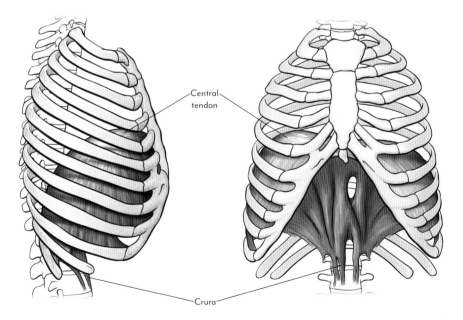

Central tendon

Crura

Fig. 7-8. The diaphragm.

THE DIAPHRAGM

The *diaphragm*, which means "partition wall," forms a boundary between the lungs and heart above and the abdominal contents below (Fig. 7-9). When the diaphragm contracts, it presses down upon the abdominal contents, distending the belly.

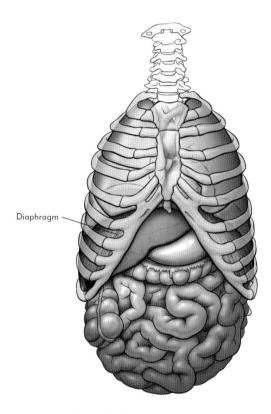

Diaphragm

Fig. 7-9. The diaphragm is a "partition wall" that forms a boundary between the lungs and heart above, and the abdominal contents below.

Breathing and Upright Support

Breathing is based on the movement of the ribs and the diaphragm, but these elements are dependent on the dynamic, flexible working of the upright system (Fig. 7-10). If this system is not working efficiently, the ribs become distorted and lose flexibility, thoracic capacity is impaired and, to put it in colloquial terms, we hold our breath. Correcting this is not simply a matter of practicing breathing exercises, which cannot reinstate the proper working of the system upon which breathing depends. To understand breathing, we must understand its dependence on the healthful working of the overall musculoskeletal system. In short, the upright system is the key to full, healthful breathing—a fact that has not yet been recognized by fitness experts, bodyworkers, massage therapists, or the medical sciences in general.

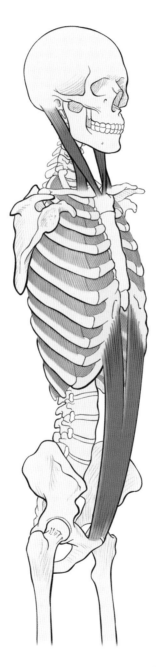

*Fig. 7-10. Breathing is directly
related to our upright support.*

THE RIB CAGE AND "WIDENING" THE BACK

When the trunk is fully lengthened and supported, the rib cage can fully open up, particularly in front, where it tends to shorten. The rib cage as a whole then has its full volume and flexibility (Fig. 7-11a).

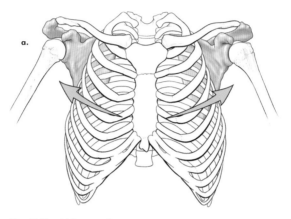

Fig. 7-11a. When widening occurs across the shoulders, the rib cage can open up.

The pectoralis major and latissimus dorsi muscles attach into the upper part of the humerus, allowing the shoulders to widen and release and the arm to widen. This widening of the shoulders removes the downward pressure of the shoulder girdle (and arms) on the rib cage and allows the trunk and ribs to open and release in front (Fig. 7-11b).

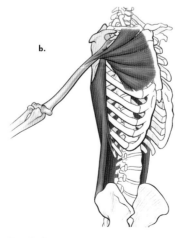

Fig. 7-11b. Widening depends on key muscles naturally releasing into length.

When the trunk is lengthening and the back is full and supportive, the thoracic region becomes more flexible and open (Fig. 7-11c). The costal joints can then move and the ribs can fully widen apart, as if a window is opening in the middle back. This allows for full and unimpeded breathing, which comes, not from the diaphragm or the movement of the ribs in front alone, but from the back, where the back needs to be full and supportive to allow the ribs to move and the joints to function freely (Fig. 7-11d).

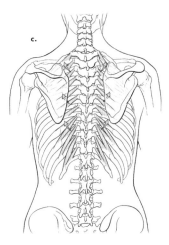

Fig. 7-11c. Widening in back allows the costal joints to work efficiently (blue arrows).

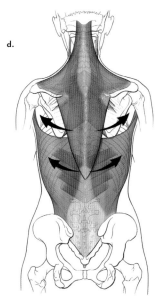

Fig. 7-11d. When the whole back is toned up and widening, the ribs can move freely.

BREATHING AND UPRIGHT POSTURE

In contrast to the ribs of a four-footed animal, which hang freely below the spine, the ribs in humans sit in front of the vertically placed spine. Because gravity is pulling them downward, the ribs often become collapsed and limited in their movements. This is why to function properly, the rib cage and breathing depend upon the lengthened support of the trunk, including lengthening of the flexors in front.

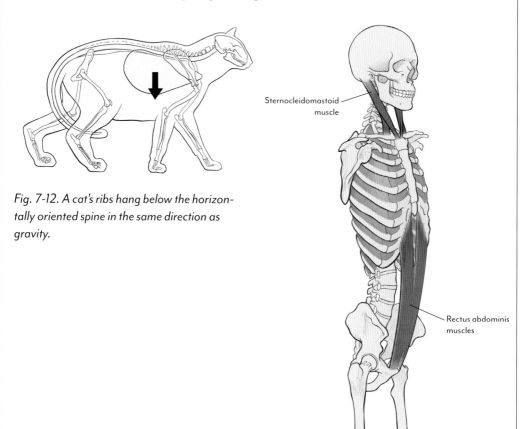

Fig. 7-12. A cat's ribs hang below the horizontally oriented spine in the same direction as gravity.

Sternocleidomastoid muscle

Rectus abdominis muscles

Fig. 7-13. Flexor support of sternocleidomastoid muscle and rectus abdominis muscles.

THE FREEDOM OF THE DIAPHRAGM DEPENDS ON THE PROPER WORKING OF THE UPRIGHT SYSTEM

We've seen that, when we breathe in, the diaphragm contracts and descends; to breathe out, it goes back to its domed position. If you habitually hold your breath, this action is impeded, which is why it is useful to focus on the out breath. But the action of the diaphragm, which broadly attaches to the ribs in front, is intimately connected to the rib cage and trunk as a whole (Fig. 7-14). When we shorten in stature and fix the ribs, the diaphragm becomes held and, in spite of its seeming independence, its full excursion is impeded. When the system as a whole lengthens and the ribs let go, the diaphragm can fully release to its domed position; the breath is then allowed to go out fully, and the breathing becomes deeper.

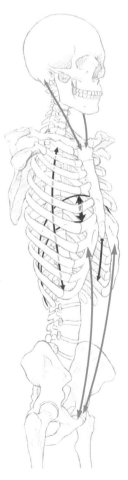

Fig. 7-14. The free movement (thick blue arrow) of the diaphragm (indicated by the dark blue horizontal lines showing the raised and lowered positions of the diaphragm) depends on a lengthening spine (vertical blue line), together with the flexor support of the trunk (turquoise lines), and the support of the rib cage (thin vertical blue line with arrows).

THE DIAPHRAGM AND THE RIBS

When the diaphragm contracts, the central tendon is drawn downward and the circumference of the diaphragm, where it attaches to the ribs, is pulled upward (Fig. 7-15). During inhalation, the lower ribs are thus raised by the diaphragm. If, as in the case of many singers, one cultivates the habit of taking breath or uses excessive effort to try to "support" the breath, this can end up overworking the diaphragm, which becomes chronically contracted. This has the effect over time of distending the ribs and distorting the lower rib cage.

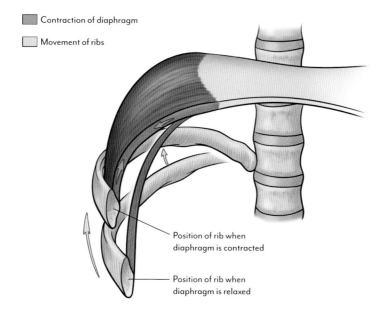

☐ Contraction of diaphragm

☐ Movement of ribs

Position of rib when diaphragm is contracted

Position of rib when diaphragm is relaxed

Fig. 7-15. The action of the diaphragm and the ribs.

HOW WE BREATHE

Breathing is the process of getting air into and out of our lungs—at least, this is how we often speak about it (we "take a breath," we "fill our lungs with breath," we "gasp for air"). But breathing is not really about what we do with the air, it is about what we do with ourselves. To take air in, we must increase the space within the thoracic cavity, which we do by moving the ribs and lowering the diaphragm. The air flows not because we do something to the air but because we do something with ourselves.

This process is roughly analogous to getting air into a smoke-filled room. When we open the doors or windows, we allow fresh air to fill the room. When we do this, air comes in of its own accord, just as, when we increase the space inside the thorax, air comes into our lungs to fill the space. When the space gets smaller, the air, which at this point is actually not the same as the air we breathed in but contains the waste products of used-up air, is forced out. When we breathe, we don't simply open the door but do something more like expanding the entire room; this causes air from the outside to flow through the doorway into the lungs. The point is that, when we breathe, we don't do something to the air but to the space around it.

When we speak of taking a breath, then, we are not being quite accurate. We don't breathe by doing something to the air but by moving the ribs and diaphragm, which increases and decreases the space within the thorax so that air rushes in and out. It is, of course, perfectly acceptable to say that we *take* or *get* air, but it reflects the very mistaken notion that, to breathe well, we should think about breathing. Nothing could be further from the truth. When we think about breathing, we do all the things that interfere with the movements of breathing: we lift the chest, we raise the ribs to suck in air, we forcibly contract the diaphragm. To ensure good breathing, we have to stop thinking about these things and, instead, focus on the bodily coordinations that will ensure healthful breathing. This principle is particularly important for singers, who are often so preoccupied with getting air into their lungs that they end up doing the very things that interfere with breathing. We obviously need air to speak or sing, but concern about breathing elicits the very habits that interfere with the movements on which breathing is based. The best way for a singer to breathe is to stop thinking about getting air in and, instead, to take care of the bodily coordinations that ensure healthful breathing.

This also explains why methods for freeing or releasing the breath fail to address what breathing is really about. To improve breathing, we must improve the bodily coordination on which it depends, and thinking about breathing does the opposite. Breathing exercises may temporarily improve breathing or invigorate the organism by increasing oxygen intake. But ultimately, they can only interfere with the processes upon which breathing depends because, whatever specific results are achieved, the act of performing breathing exercises brings into play the very tensions that disturb the natural movements of breathing.

EXERCISE:
The Controlled Exhalation and the Whispered "Ah"

Almost all of us, when we speak or do things, tend to hold our breath without noticing that we are doing so. This habit seems innocuous enough, but it is actually a form of tension that interferes with the way the body is naturally designed to work. We have to learn to leave the breathing mechanism alone by a) not taking breath and learning to allow air to come into the lungs without trying to take a breath, and b) not collapsing or creating pressure when the air goes out or when we vocalize. A very effective way to do this is to extend or control the exhalation.

It may seem paradoxical to suggest that, to breathe better, we should stop taking breath in and instead focus on letting it out. But breathing takes place as a result of flexible movements of the diaphragm and ribs, which is what makes it possible to breathe in and out. If we are holding our breath all the time, these movements will be interfered with. We have to learn to stop holding our breath if we want to breathe in fully.

A simple way of understanding the controlled exhalation is to think of it in terms of not holding the breath. If you lift a heavy object or prepare to speak, you'll notice that the initial response is to stiffen and hold the breath. If, at the moment you begin to hold the breath, you stop and let the breath out slowly through your lips or teeth, you are performing a controlled exhalation. Controlling the exhalation in this way allows us to shift our attention away from holding or taking breath (which is the universal tendency) and instead to focus on letting the breath out—which is to say, not stiffening and holding the breath. This not only frees the diaphragm and ribs but also has a calming effect on the nervous system.

Exercise 1: The Controlled Exhalation

There are several ways of performing a controlled exhalation: by blowing through the lips, hissing through the teeth, breathing out slowly through the nostrils (as if you are making an "ng" sound), and whispering. In the following exercise, feel free to experiment with different ways of performing the controlled exhalation, and do what feels most comfortable.

1. Sit comfortably in a chair, back supported by the back of the chair and feet flat on the floor. Notice the breath going in and out through your nostrils. Remember that we should not normally be breathing through the mouth, and that breathing freely through our nostrils is a good way to ensure that we are not interfering with our breathing.

2. Without altering your breathing in any way, make a whooshing or "sss" sound for several out-breaths. Don't take in air to do this; simply let it come in on its own, and make sure you are breathing through the nostrils on the in-breath.

3. Finally, be sure not to turn the controlled exhalation into a sigh by collapsing during exhalation, which may seem relaxed but is in fact collapsing the chest and has little to do with coordinated breathing or vocal production. A true controlled exhalation can be brought about only when one has the clear intention to prolong the out-breath in a steady way and maintains a coordinated, energized state of the muscular system while doing so.

Exercise 2: The Whispered "Ah"

A particularly useful way of controlling the exhalation is to whisper. When we whisper, we are controlling or extending the exhalation, but in a way that is directly related to using the larynx, or vocalizing.

If you ask the average person to whisper an "ah" sound, you hear a somewhat throaty "huh" sound. In contrast, if you are clear that the vowel sound you want to produce is not an "uh" but an "ah," the exhalation is lighter and the throat is more open, which is why the whispered "ah" uses an "ah" in the first place.

1. With your feet flat on the floor, and sitting comfortably in the chair, notice the breath going in and out through your nostrils. Remember that we should not normally be breathing through the mouth, and that breathing freely through our nostrils is a good way to ensure that we are not interfering with our breathing.

2. Without altering your breathing in any way, open your mouth and produce an "ah" sound by whispering. Remember that the vowel sound is "ah" and should sound open, like when the doctor puts the wooden depressor on your tongue and asks you to say "ah." It is also prolonged, so that it is obvious that you are clearly intending to produce the sound.

3. Do several of these "ah" exhales in a row, allowing the air to come in by itself without actively taking the breath.

When doing this exercise, make sure you are breathing through your nostrils on the in-breath. During normal breathing and speech, we tend to gasp in air through the mouth. Allowing air to come in through the nose by consciously preventing that tendency tends to restore a normal breathing pattern. Whenever we take air in, it should come in through the nose as a rule, even if we are doing a controlled exhalation or speaking out through the mouth.

WHY WE BREATHE

When we breathe, we take air into the lungs by expanding the rib cage and lowering the diaphragm; we breathe out when the ribs descend and the diaphragm ascends. For most of us, breathing refers precisely to this flow of air into and out of the lungs. We "breathe deeply," we "fill our lungs," we perform deep breathing exercises designed to increase the flow of air into our lungs.

But breathing isn't just about getting air into our lungs. For our cells to function, oxygen is needed to break food down so that it can be converted into energy-storing molecules known as *adenosine triphosphate (ATP)*. As an essential fuel for maintaining cellular activity, we must constantly produce ATP; without a constant supply of oxygen, we cannot make ATP, and the cell dies. This is why we cannot live for more than a few minutes without oxygen: the heart stops, brain cells begin to die, and death soon follows.

This is where breathing comes in. Our blood carries oxygen molecules to our tissues, and our lungs are where we oxygenate the blood. When we take air into our lungs, blood flowing though the fine capillaries in our lungs absorbs the oxygen. The oxygen-rich blood is then carried to tissues throughout the body, where it is absorbed into the cells. At the same time, waste products from the cells are transferred to the blood so that, when it flows back to the lungs, these waste products can be exhaled in the form of carbon dioxide.

Respiration, then, has two phases. In the first phase—sometimes called *external respiration* because it involves direct contact with the outside—we take air into our lungs. The alveoli, the little sacs of the lungs, intersect with fine capillaries and here, where the oxygen meets the capillary walls, the blood absorbs the oxygen and circulates it through the body. In the second phase—called *internal respiration* because it takes place deep within cells—the oxygen is transferred to tissues throughout the body including the brain, the heart, skeletal muscles, and organs. We think breathing goes on in the lungs, but it really happens at the cellular level throughout our body's tissues (Fig. 7-16).

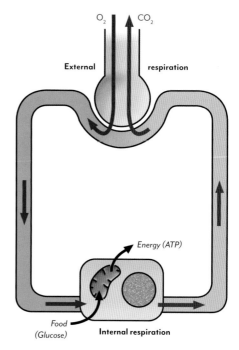

Fig. 7-16. Internal and external respiration.

Neither of these phases, however, can take place without bodily movement. We get air into the lungs by alternately expanding and contracting the thorax and contracting the diaphragm, and these movements are just that: movements. Breathing happens at the cellular level, but it depends on the flexible movements of the ribs and diaphragm, which are in turn part of the upright support system. If we interfere with these movements because we are habitually holding our breath, trying to take deep breaths, or mouth-breathing, we interfere with both the movements and the upright support system. The best way to see to it that our tissues are receiving the oxygen they need is to stop holding our breath and to breathe quietly and slowly through our nostrils so that our lungs receive more air and waste products can be efficiently expelled. If, in addition, we learn to let the air out slowly by controlling the exhalation, then—without even thinking about breathing—the air can come in on its own and we are able to breathe easily and efficiently.

TONGUE, PALATE, AND MOUTH BREATHING: BREATHING THROUGH YOUR NOSTRILS

As you sit here reading, are you breathing through your nostrils or your mouth? Do you find yourself mouth-breathing when you are working at a desk, or when you wake up in the morning? Although we sometimes need to breathe through our mouths, as a general rule, we should breathe not through our mouths but through our nostrils. Mouth-breathing is a collapsed form of respiration that is associated with shallow breathing, collapsing the palate, and slumping. Like being sedentary, mouth-breathing has become something of a health hazard for young and old alike, which is why practitioners of various health practices urge us not to breathe through the mouth and tout the benefits of nasal breathing.

But breathing through our nostrils is not simply beneficial, as if nasal breathing is a trick for making you healthier. Many reptiles and all mammals possess a secondary palate that separates the digestive and respiratory passageways, making it possible to take food or water while continuing to breathe (Fig. 7-17). This separation is especially important for herbivorous mammals, who are able to detect predators while feeding, as well as for suckling infants, who breastfeed for long periods and must be able to continue to breathe while doing so. (It is perhaps for this reason that infants are considered "obligate nasal breathers"—that is, they can breathe through their mouths but prefer nasal breathing) (Fig. 7-18).

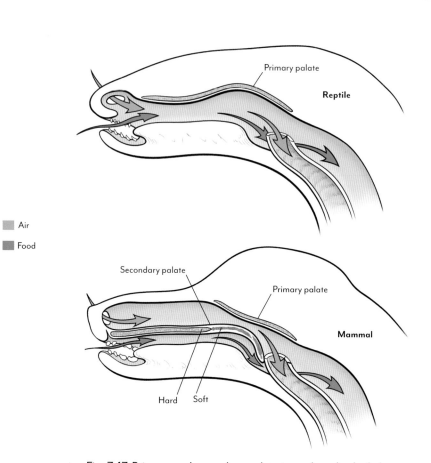

Air

Food

Primary palate

Reptile

Secondary palate

Primary palate

Mammal

Hard   Soft

Fig. 7-17. Primary and secondary palate in: a. A snake; b. A dog.

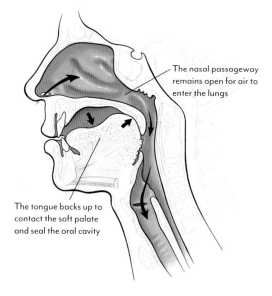

The nasal passageway remains open for air to enter the lungs

The tongue backs up to contact the soft palate and seal the oral cavity

Fig. 7-18. Our secondary palate makes it possible to breathe through our nostrils while chewing food.

As a dedicated pathway for breathing, the nasal passages are specially adapted to filter out foreign particles and to moisten and warm the air. Breathing through the nostrils also slows down the flow of air into the lungs, which enhances oxygenation of the blood. Mouth breathing tends to be shallow and quick; when we take in air through our nostrils, we breathe more fully and engage more of the lungs. Breathing through the nasal passages also releases nitric oxide molecules into the inhaled air, which act as "airborne messengers" in the cardiovascular system that, among other things, regulate blood pressure, increase oxygen uptake in the lungs, and boost immune function. All these benefits, which are part of how we are inherently designed to breathe, are lost when we breathe through our mouths.

When you breathe through your nostrils, your mouth should be closed with your tongue naturally arching in your mouth and contacting the hard palate (Fig. 7-19). Breathing freely through your nostrils, with your tongue in this "resting position," is a good way to ensure that you are not holding your breath.

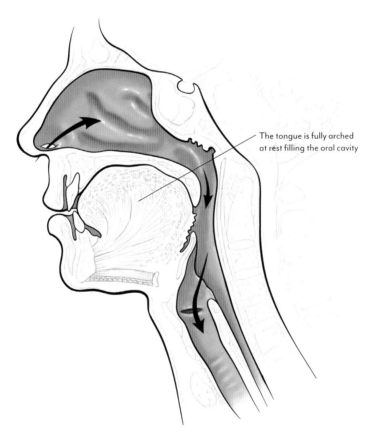

The tongue is fully arched at rest filling the oral cavity

Fig. 7-19. When you breathe through your nostrils, your mouth should be closed with your tongue naturally arching in your mouth and contacting the hard palate.

EXERCISE:
Not Gasping in Air When We Speak

Although we need air to speak or sing, it does not follow that, to get the air we need, we should actively take a breath. Many of us—even trained singers—tend to gasp in breath when we speak. This tendency, which is often cultivated to a very harmful degree by voice users, is associated with pulling back the head, raising the chest and shortening in stature. All these tendencies interfere with breathing and vocalizing, which is why learning to speak or vocalize without taking breath is a crucial skill for voice users and fundamental to the art of effortless vocalization.

One way to break the habit of taking breath to speak is to lie down or sit quietly and to breathe normally through your nostrils with your mouth closed and your tongue in its natural resting position—that is, arched so that the body of the tongue contacts your hard palate. When you are aware of the breath going in and out through the nostrils, this ensures that you are leaving your breathing alone. Taking this as a starting point, and using the lines of a poem you have memorized, try the following exercises.

Exercise 1: Non-Doing and Speech Movements

1. Spend a few minutes lying down in the semi-supine position with some books under your head, making sure that your neck is free and your back is fully supported on the ground (see pages 59 and 83).

2. Notice the breath going in and out through your nostrils. Make sure your lips are together and your tongue is naturally arching in your mouth.

3. When you're satisfied that you are leaving the system alone nicely, see if you can open your jaw while continuing to breathe through your nostrils. Notice if you have stiffened your neck muscles, tightened your throat, or held your breath. Opening the jaw is, of course, one of the elements of speech and is therefore closely associated with the basic pattern of misuse; learning to open the jaw without interfering with your head balance and without disturbing the flow of breath in and out of the nostrils is critical to coordinated speech.

4. When you can move your jaw comfortably without interfering with the flow of air through your nostrils, try moving your tongue, lips, and the muscles of your face. If you can do this without interfering with your breathing and with your overall muscular system, you are well on your way to being able to speak in a coordinated fashion and can begin to apply this in everyday speech.

Exercise 2: Inhibition and Nostril Breathing

1. Continue to notice the breath going in and out through your nostrils.

2. Without interfering with this natural flow of air, speak one line and then close your mouth and let your lips come together so that the breath comes in through your nostrils.

3. Allowing the air to continue to go in and out through your nostrils, speak the second line. Notice if there is any change in the quality of sound you produce when you do. Let your lips come together.

4. See if you can recite three or four lines of your poem in this way—that is, allow the breath to go in and out through your nostrils for a few breaths before reciting each line, and open your mouth only to speak. Notice the change in the quality of your voice when you do this.

If you can speak without taking breath between phrases, try reciting several lines of a poem, seeing to it that, whenever you begin a new line, you close your mouth so that the air enters your nostrils and doesn't come in through your mouth. This is an invaluable exercise for preventing unnecessary interference with the breathing system and with the throat while speaking, and it provides a crucial foundation and point of reference for becoming aware of unnecessary and harmful habits in normal, everyday speech.

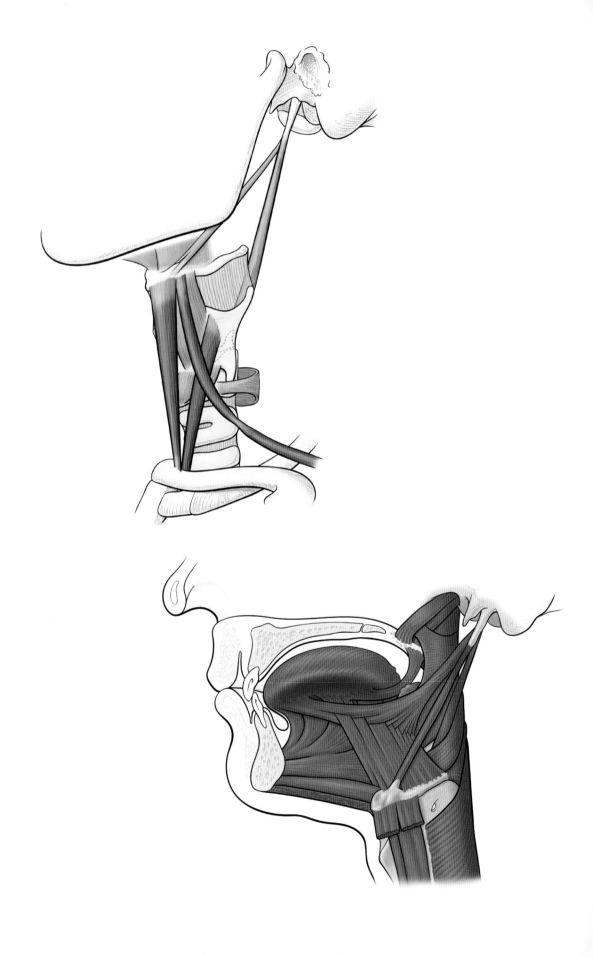

The Larynx and Throat

8

The function of the throat is to provide a pathway for food, water, and air. All the key structures of this region—pharynx, jaw, tongue, and larynx—are acted upon by a series of muscles that assist in feeding, chewing, swallowing, and breathing. In addition, the larynx is a sound-producing organ, and many of the structures in this region contribute to speech production.

The Key to the Larynx and Throat

Although the throat contributes mainly to breathing, feeding, and vocalization (and only secondarily to posture and movement), all the key structures in this region are directly or indirectly suspended from the base of the skull and acted upon by a network of muscles attaching to the jaw, skull, and sternum. Key among these structures is the hyoid bone, which acts as a central hub of support for the tongue and larynx and which is suspended within a network of muscles. When working freely, the muscles attaching to the hyoid bone must release to allow the throat to become freely supported within its suspensory network.

Contraindications

When we use the voice in such a way that we tighten the throat and depress the larynx, the pharynx becomes narrowed. Coupled with the tendency to raise the larynx and tighten the palate during speech, the entire throat region becomes fixed, dragging down upon on the skull and contributing to a general shortening of stature.

WHEN THE THROAT IS TIGHT

1. The larynx is depressed, with a corresponding shortening of stature.
2. The hyoid bone is fixed.
3. The throat is constricted and the larynx is raised in speaking.
4. The soft palate loses flexibility.
5. The jaw is tight and the muscles on the underside of the tongue are overworked.

WHEN THE THROAT IS FREE

1. The depressors of the larynx (the strap muscles) are released.
2. Front length is restored.
3. The larynx is neither raised nor depressed.
4. The throat is open and the soft palate is flexible.
5. The hyoid bone and tongue release within their muscular scaffolding.
6. The jaw is free.

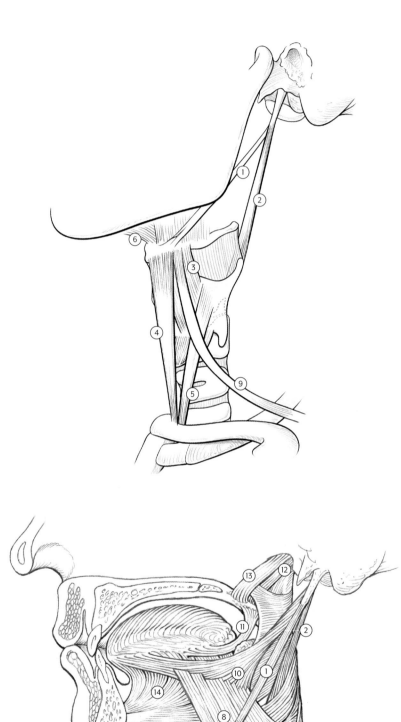

ORIGIN	MUSCLE	INSERTION
Styloid process	1. Stylohyoid	Hyoid bone
Mastoid process	Digastricus	Jaw
Styloid process	2. Stylopharyngeus	Thyroid cartilage/pharynx
Thyroid cartilage	3. Thyrohyoid	Hyoid bone
Sternum	4. Sternohyoid	Hyoid bone
Sternum	5. Sternothyroid	Thyroid cartilage
Jaw	6. Mylohyoid	Hyoid bone
Jaw	7. Geniohyoid	Hyoid bone
Hyoid bone	8. Hyoglossus	
Scapula	9. Omohyoid	Hyoid bone
Styloid process	10. Styloglossus	Tongue
Soft palate	Palatopharyngeus	Thyroid cartilage
Soft palate	11. Palatoglossus	Tongue
Temporal bone	12. Levator veli palatini	Soft palate
Sphenoid bone	13. Tensor veli palatini	Soft palate
Mandible	14. Genioglossus	Hyoid bone/tongue

The Suspensory Muscles of the Larynx

The muscles of the larynx, throat, and jaw play a central role in our upright design. In Chapter 3 we saw that the rib cage and innards create a drag on the front of the body, contributing to the tendency to lose length in front. Because the larynx and tongue are essentially suspended from the skull, they too—if used improperly—can pull down in front, dragging upon the head, jaw, and upper spinal column. If the tongue and larynx are depressed in this way, the upper spinal column and trunk cannot fully lengthen and the head cannot balance freely upon the spine. Thus, a freely suspended throat that does not drag on the skull is an essential component of our ability to lengthen against gravity.

The Flexor Sheet of the Throat

The throat muscles form an intricate network or webbing on the neck and underside of the jaw (Figs. 8-1 and 8-2). In addition to moving the larynx and helping to form speech sounds, these muscles assist in such basic life functions as eating, breathing, and swallowing.

Although the throat is the most complex network of muscles in the body, this critical system is often omitted from books on the anatomy of movement. But the jaw, larynx, tongue, and throat are all suspended from the skull and, in this sense, are all integral to our upright support system and profoundly influence the working of the musculoskeletal system as a whole.[1] Over time, many of us use the voice in such a way that we collapse and tighten the throat. The musculature of the throat becomes constricted and the larynx is pulled downward by overworked muscles, creating a drag on the skull, compromising the length of the trunk, and interfering with the upright support system.

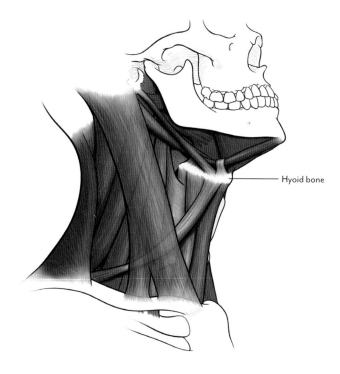

Hyoid bone

Fig. 8-1. An intricate network of throat muscles is involved in eating, breathing, and speaking.

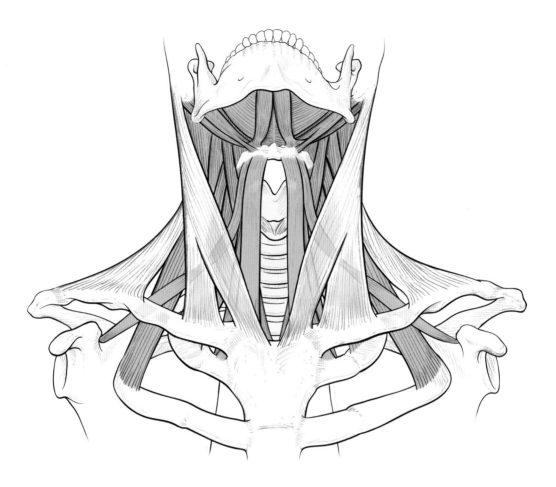

Fig. 8-2. The deeper muscles of the throat form a flexor sheet linking the jaw, larynx, and rib cage.

A sedentary lifestyle can also contribute to the tendency of the palate, tongue, and throat to collapse. While working at a desk, many people collapse the palate and, instead of breathing freely through the nostrils, breathe heavily through the mouth. Even while sleeping, many of us collapse and tighten the throat muscles in such a way that they drag upon the head and spine.

When the throat musculature releases, the downward drag on the skull is eased and the structures of the throat, including the larynx within its muscular scaffolding, become properly suspended from above. This results in a freer, more open use of the voice while, at the same time, allowing the body to regain its optimal length.

THE LARYNX AND THE THROAT

The larynx forms a sphincter whose most basic function is to protect the airway (Fig. 8-3).[2] As such, the larynx is not an isolated and discrete organ but evolved in conjunction with breathing and works cooperatively with the respiratory system to produce sound. It is also supported by a network of muscles that pull upon it from different directions to assist the stretching of the vocal folds in creating particular pitches and qualities of sound (Fig. 8-4).

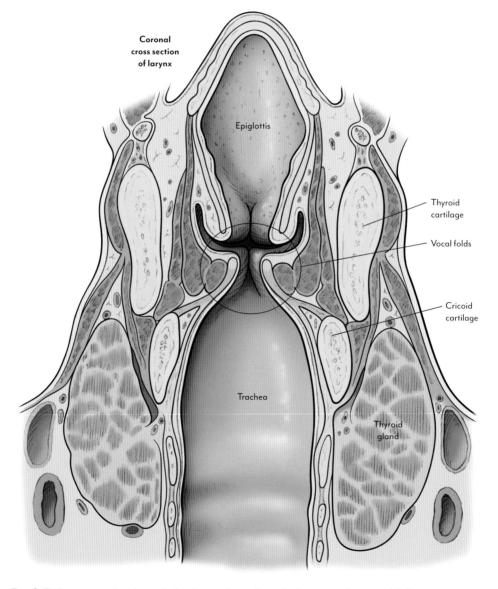

Fig. 8-3. A cross-section through the throat shows how the larynx and vocal folds function as a sphincter muscle.

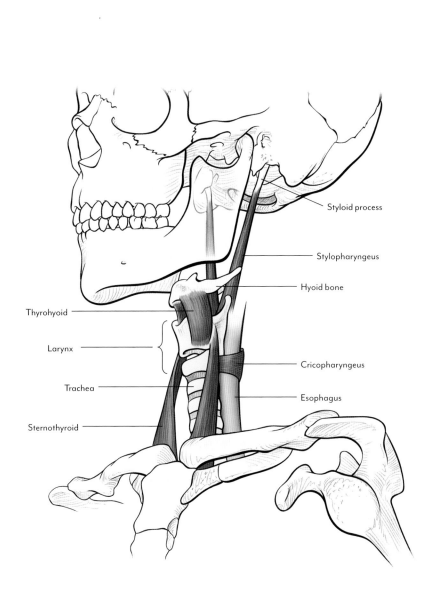

Fig. 8-4. The suspensory apparatus of the throat.

THE SUSPENSORY MUSCLES OF THE LARYNX

The musculature of the throat is a crucial part of the singing and speaking apparatus. To form speech sounds, dozens of muscles are brought into play. To speak and sing, these muscles antagonistically pull on the larynx from different directions, creating the necessary support for it to function optimally (Fig. 8-5).[3] If we represent the muscles as guy wires, we see a tensegrity-type structure emerging, such as the ones we looked at in Chapter 5.

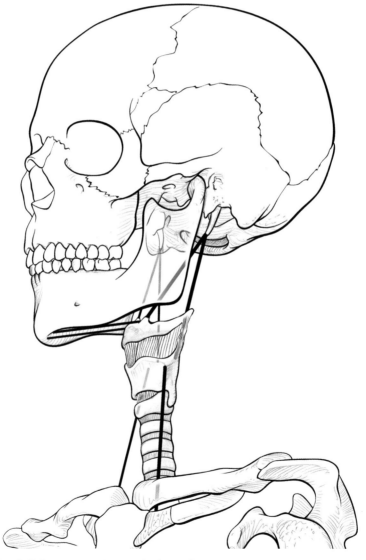

Fig. 8-5. Represented here in the form of guy wires, we can see that the muscles of the throat act as a tensegrity structure to support the hyoid bone and larynx.

The Base of the Skull and the Suspended Throat

On the base of the skull, set in from the sides, are two small but prominent spikes called the *styloid processes* (Fig. 8-6a and b). It is from these two key attachments that the hyoid bone and larynx are suspended from the skull. The muscles of the throat, or pharynx, also hang from the skull just in front of the *foramen magnum*, the hole in the base of the skull through which the spinal cord extends down the spine (Fig. 8-6b and c), so that virtually all the structures of the throat hang from the skull and are, in this sense, dependent on—and profoundly influence—the workings of the upright support system.

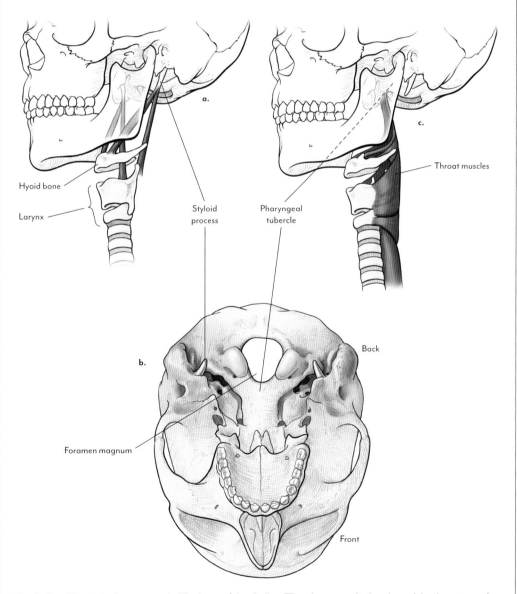

Fig. 8-6. a. The styloid processes; b. The base of the skull; c. The pharyngeal tubercle and the throat muscles.

THE DEPRESSED LARYNX

Over time, many of us use the voice in such a way that we collapse and tighten the throat. The musculature of the throat becomes constricted and the larynx is pulled downward, creating a drag on the skull and the upper spinal column and compromising the length of the trunk (Fig. 8-7 and Fig. 8-8).[4] In the boy pictured below (Fig. 8-8), we observe shortened neck muscles, with the head tipping backward in relation to the spine, and compression in the area of the larynx. This results in loss of postural tone, and excessive curves, as we saw previously in Fig. 4-6 (page 76).[5]

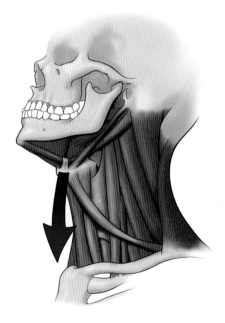

Fig. 8-7. The depressed larynx.

Fig. 8-8. Faulty action in the throat and tongue region can affect total postural support.

THE FREE THROAT

The jaw, larynx, and throat are all suspended from the skull. If we shorten and collapse this musculature, the muscular webbing of the larynx and throat tighten and drag downward; the throat itself becomes collapsed and closed (Fig. 8-9a). When the head balances forward and up, as it is naturally designed to do, this network of throat muscles becomes more open, and the larynx, which is supported within the suspensory muscles, is able to function optimally. When the musculature of the throat releases, the larynx functions more freely and we hear a "ring" in the voice, indicating that the throat is more open and that the speaking voice is activated and produces sound with less effort (Fig. 8-9b).

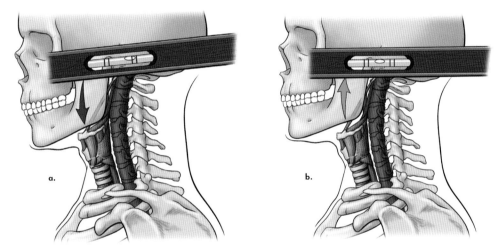

Fig. 8-9. a. Tightening in the muscles of the throat leads to drag on the skull and shortening in the neck; b. When the throat musculature releases, the downward drag on the skull is eased and the structures of the throat, including the larynx within its muscular scaffolding, become properly suspended from above.

THE SOFT PALATE: THE SECOND DIAPHRAGM

The soft palate forms a kind of movable flap or curtain that separates the nasal and oral cavities. Sometimes called the *velum* (Latin for "curtain" or "veil"), the soft palate functions as a diaphragm that acts as a valve to close off these cavities as needed. While working at a desk, many of us collapse the palate and, instead of breathing freely through the nostrils, breathe heavily through the mouth. This affects not only the immediate structures of the throat but vocal use, breathing, and muscle tone in general. To maintain the health of the voice and throat, as well as our overall muscular system, the palate must retain a certain degree of flexibility and tone.

Viewed from the side, the soft palate can be seen as an extension of the hard palate that connects the oral cavity with the pharynx and nasal passageways (Fig. 8-10). When the soft palate closes against the back of the tongue, this creates an oral seal that makes it possible to breathe through the nostrils while we chew food or swish water in the mouth (Fig. 8-11a). When the soft palate contacts the back of the pharynx, this creates a nasal seal (Fig. 8-11b). And when it contacts both the tongue and pharynx there is total closure of the nasal and oral passageways, as when we swim underwater (Fig. 8-11c).

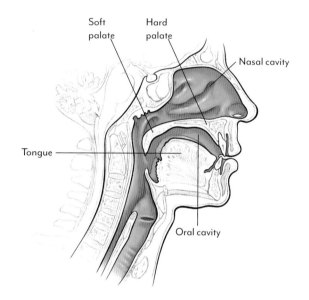

Fig. 8-10. The soft palate forms a diaphragm separating the nasal and oral cavities.

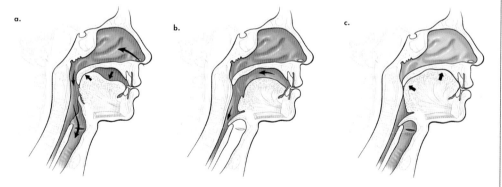

Fig. 8-11. a. Oral seal—tongue backs up against the soft palate (short arrows) letting air come in through the nose (long arrows); b. Nasal seal—soft palate contacts the back of the throat during swallowing (arrows); c. Total closure of the throat with the tongue filling the oral cavity (arrows).

The movements of the soft palate are intimately involved in swallowing and breathing. If you look directly into the oral cavity, you will see arches running down the sides of the pharynx—two on each side (Fig. 8-12). These arches are actually muscles: the first, palatoglossus, attaches to the tongue; the second, palatopharyngeus, to the larynx. When we swallow, these muscles contract, depressing the palate and, at the same time, raising the tongue and larynx (Fig. 8-13a). Muscles attaching to the top of the soft palate perform the opposite function, elevating the palate and raising the arches (Fig. 8-13b). So the palate can make two basic movements. When we swallow, the larynx and soft palate move toward each other; when we yawn or take a deep breath, they move away from each other.

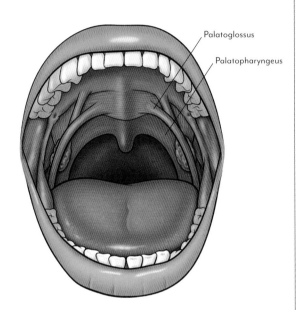

Palatoglossus

Palatopharyngeus

Fig. 8-12. The arches of the palate.

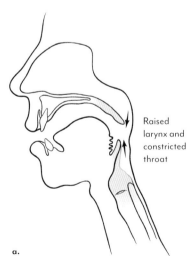

Raised larynx and constricted throat

a.

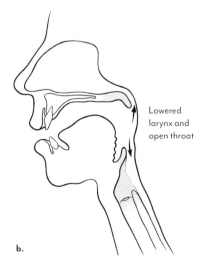

Lowered larynx and open throat

b.

Fig. 8-13. The palate and larynx work in opposition.

These actions are quite normal but, with our harmful, sedentary lifestyles, we tend to collapse the soft palate, interfering with its natural flexibility and tone. When this happens, it can no longer arch fully and it "closes" the throat. When we yawn, we are activating these muscles in an antagonistic way that allows them to stretch, as we do with the entire body when we have a good stretch in the morning—a process that can be observed in cats and young infants, who wake from a nap and "yawn" with the entire body and with the throat as well. This tends to stretch and tone the muscles and to prepare them for activity. Yawning also helps to restore stretch and tone to the throat muscles if they have become tight from long hours of using the speaking voice. Humming is another useful exercise for raising and toning the palate; as you'll see in the following exercise, with just a bit of experimentation, you can learn to consciously raise the palate and open the throat. The soft palate is thus a second diaphragm that must remain flexible and toned, even into old age.

EXERCISE:
Opening Your Throat

To experiment with arching the palate and opening the throat, begin by humming a melody and, then, without changing the hum, open your mouth. When you do this, the "m" sound of the hum will turn into an open vowel, and you may find that your throat is more "open" and the palate is more arched and raised than if you simply start singing with your mouth open, in which case the throat is more closed (Fig. 8-14). Learning to tone and arch the palate in this way is an essential component of virtually all vocal techniques.

You can also try sneering, or imagine smelling a rose. Let the toning of your cheeks and eyes extend into your throat to help lift and tone your palate.

Another way of opening the throat is by yawning, which tends to raise and arch the palate. As we saw earlier, yawning is nature's way of stretching and toning the throat if we have been sleeping or sedentary. Because our throats tend to become habitually constricted, it is useful to experiment with this. However, the yawn should not be forced or overdone but should be allowed to come naturally.

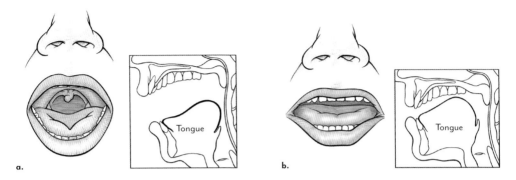

Fig. 8-14. Tongue, palate, and jaw positioning with: a. An open throat; b. A closed throat.

Finally, it is useful to listen to singers and voice-users who sing with an open throat. Hearing good operatic and classical singing attunes the "ear" to good vocal use and makes it easier to distinguish free and open-throated singing from closed and strained vocal use.

After you become familiar with these procedures, it is useful to experiment with how they affect vocal tone. Many people have a throaty and heavy way of speaking that is connected with being serious or worried; with practice, it becomes easy to identify this and to learn to develop better placement and tone by thinking of the eyes, cheeks, and nostrils. At first, you might feel vulnerable when singing or speaking in this way, but with experimentation it gets easier and even fun. With practice, paying attention to the facial muscles can have a profound and immediate effect on vocal quality and production. After a while, these exercises for the face and throat become second nature and can easily be put together with the other elements of the whispered "ah"—see page 155.

THE HYOID BONE

The hyoid bone is the only bone in the throat region and it serves three essential purposes.

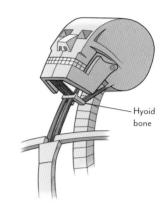

Hyoid bone

- It serves as an attachment for the larynx and for the muscles that elevate and lower the larynx in swallowing (Fig. 8-15).

- It is a key anchor point for the muscles on the floor of the mouth that open the jaw.

- It provides a stable base for the tongue. (For this reason it is sometimes referred to as the "tongue bone.")

Fig. 8-15. Mechanical drawing showing the function of the hyoid bone.

Where is your hyoid bone?

The hyoid bone is located on your throat just above your larynx (Fig. 8-16). If you swallow or wag your tongue, you will feel the hyoid bone move, along with your larynx. To find the hyoid bone with your fingers, place your thumb and forefinger under your jaw and then bring them back so that you are gently pinching your throat just above the larynx. You will find this small, horseshoe-shaped bone just above the larynx, which you can gently pinch between thumb and forefinger.

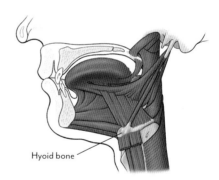

Hyoid bone

Fig. 8-16. Locating the hyoid bone.

Assessing the Throat

As the structural epicenter of the throat region, the hyoid bone is a key indicator of the condition of your throat. When the throat musculature is tight and constricted, the hyoid bone seems either fixed or dragging downward (or both), thus affecting head balance and contributing to the general "pulling down" of the entire postural system. When, in contrast, the throat is free, you can place a hand on the front of the throat, right up against the hyoid bone; there is an absence of downward drag, and the hyoid bone seems to go up and back. The body as a whole has its full front length, and with your hand at this critical marker, you can feel the upward flow of the entire body in front, allowing the neck to be free and the head to go forward and up while the body as a whole lengthens against gravity (Fig. 8-17).

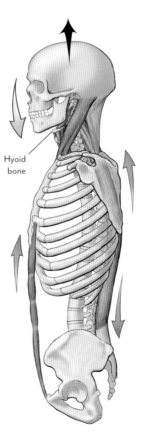

Fig. 8-17. The hyoid bone in relation to total posture.

THE JAW AND THROAT

The jaw forms a kind of movable hinge joint with the temporal bones of the skull (the temporomandibular or TM joint) and is acted upon by three sets of powerful muscles that often become chronically tense, limiting the freedom and mobility of the jaw (Fig. 8-18). It is easy to assume that this tension can be alleviated by moving the jaw or attempting to relax these muscles directly. But the jaw evolved as part of the throat and feeding mechanism; if the muscles on the underside of the jaw and throat become tight, the jaw muscles are forced to compensate by chronically tightening (Fig. 8-19). When the head balances forward on the skull and the trunk lengthens, these muscles can release and the jaw muscles, in turn, will no longer be forced to contract and will sympathetically release. For this reason, exercises for freeing the jaw muscles fail to address the underlying cause of tension in the jaw. The freedom of the jaw is entirely dependent on the throat musculature and its suspension from the skull.

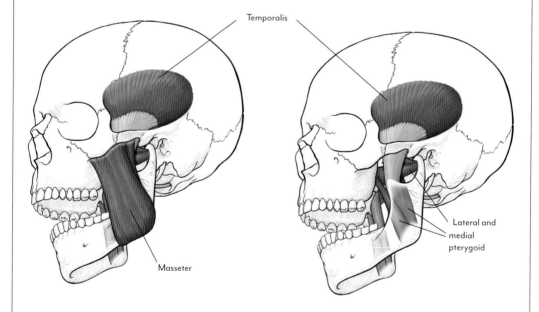

Fig. 8-18. Muscles of the jaw.

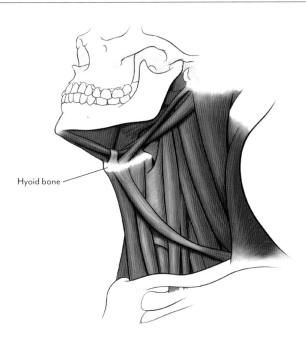

Hyoid bone

Fig. 8-19. The jaw is intimately linked with the musculature of the throat.

THE TEMPOROMANDIBULAR JOINT AND HOW IT WORKS

The jaw, or mandible, forms a joint with the temporal bones of the skull, called the temporomandibular or TM joint (Fig. 8-20). Just in front of the ear is a depression called the mandibular fossa. Sitting within this depression, the condyles of the jaw rotate to produce the hinging action of opening and closing the jaw. This articulation, however, is not a pure hinge joint; instead, the jaw "hangs" from the temporal bones, making it possible not only to hinge the jaw but also to swing it forward and back. When we open the jaw during normal speech, the jaw simply hinges (Fig. 8-21a). When we grind our teeth, or when we open the jaw more widely, the condyles not only hinge but glide forward (Fig. 8-21b). The complexity of this joint makes the jaw susceptible to stresses, as when the jaw makes clicking noises because it is not hinging efficiently, or when we grind our teeth and experience TM joint pain.

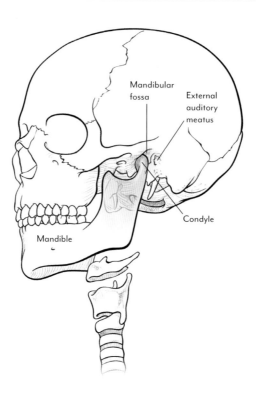

Fig. 8-20. The temporomandibular joint.

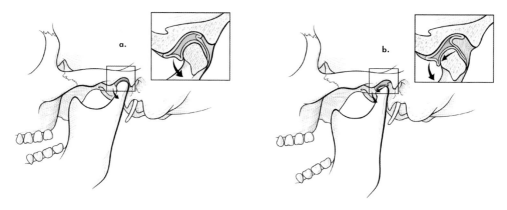

Fig. 8-21. a. Partial opening of the jaw: the jaw hinges at the joint (downward arrow) without moving forward in space; b. Wider opening of the jaw: the jaw swings or moves forward in relation to the skull (forward arrow) as it hinges to open (downward arrow).

EXERCISE:
Opening the Jaw by Gliding It Forward

If you gently open and close your jaw, you can see how the condyles of the jaw, where they articulate with the temporal bones just in front of the ears, produce the hinging action of the jaw. When you open the jaw more widely, as you do when singing, the jaw as a whole swings forward slightly. To experiment with this, open your mouth widely. Notice that this tends to retract the jaw, forcibly jamming the TM joint and preventing the jaw from gliding forward and opening freely. Now try opening your mouth again, but this time let the jaw glide forward as you open it. If you do this carefully, you will feel a pocket open just in front of your ear lobe, and the jaw will open more fully. By doing this, you have allowed the jaw to glide forward, which makes more room for the joint and allows it to open more freely.

Because most of us tighten our neck muscles and pull the head back to open the jaw, it is important to learn to open the jaw with as little tension as possible. To experiment with this, try the following exercise in semi-supine position and while sitting.

1. Open your jaw quickly and notice that it tends to retract and seems to be limited in how far it can open. You may also notice that you tightened your neck and pulled your head back to open the jaw.

2. Let your jaw close and, this time, gently slide your lower jaw forward; think of directing the tip of your tongue and your chin forward as you do so. Then gently open your jaw. You may notice that your jaw is now opened more widely than it was before.

3. Close your jaw and "direct" your tongue and chin forward as before. Then try opening your jaw once more, experimenting to see if you can do so without retracting it or pulling your head back.

4. See if you can brighten and soften your eyes while opening your jaw so that, even while you give your directions for one part of your body, you are able to think of another.

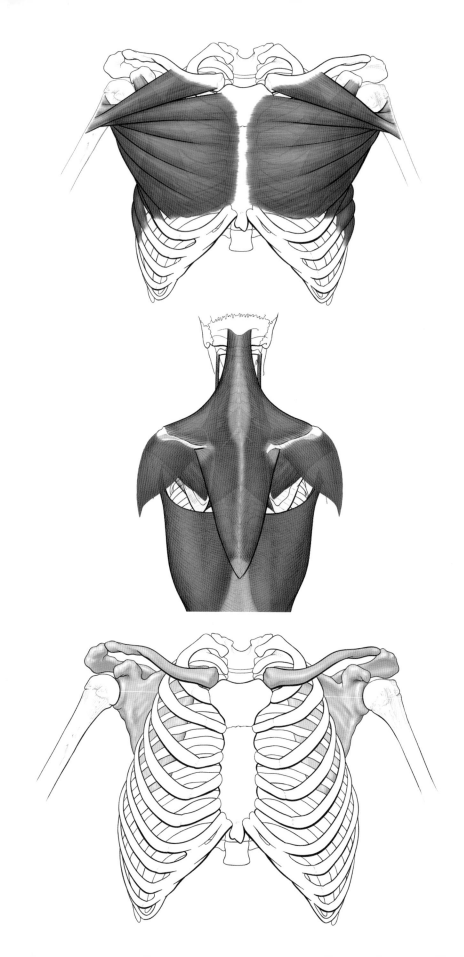

The Shoulder Girdle

9

The *shoulder girdle* is a movable cross-piece that supports the levers of the arms. It provides sockets for the shoulder joints and, because it is supported and moved by a network of muscles, it contributes to the range of motion of the arms.

Widening the Shoulders

When the flexors across the front of the shoulders are naturally lengthened, the shoulders are "widened" and linked into the back, where they get most of their support. This allows the back muscles to work in a full and supportive way and the ribs to move freely, thus restoring the mobility of the ribs and full breathing. The shoulders are thus a crucial element in restoring the widened support of the back and freeing of the rib cage.

Contraindications

Many of us, however, are constantly engaged in activities that activate the flexor muscles of the hand, forearm, and shoulders, with a corresponding narrowing or shortening across the front of the shoulders. If we slump while sitting, this further contributes to the tendency to narrow the shoulders, causing the flexors of the shoulders and arms to become habitually shortened and overworked.

WHEN WE NARROW AND RAISE THE SHOULDERS

1. The pectoral muscles shorten and fix the shoulder.
2. The rotator cuff muscles shorten and the shoulder joint becomes fixed.
3. The scapulae become fixed in back.
4. The upper arms are pulled together and put pressure on the rib cage.

WHEN THE SHOULDERS WIDEN

1. The pectoral muscles release to allow the shoulders to spread apart.
2. The scapulae muscles tone up, and the scapulae are free to move and widen apart.
3. The latissimus and pectoralis major muscles release to lend widening support to shoulders.
4. The shoulders are supported from above without being raised.
5. The rotator cuff muscles release and the shoulder joint regains mobility.

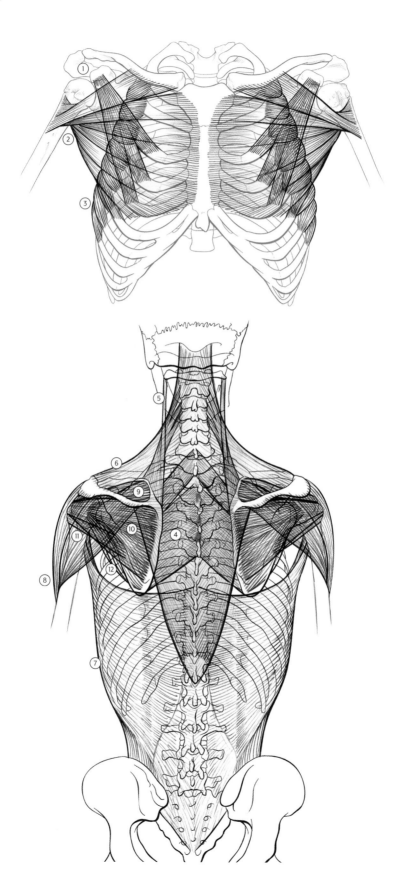

ORIGIN	MUSCLE	INSERTION
ANTERIOR SHOULDER GIRDLE		
Ribs 3–5	**1. Pectoralis minor**	Coracoid process
Medial clavicle/sternum	**2. Pectoralis major**	Upper humerus
Ribs 1–8	**3. Serratus anterior**	Medial border of scapula
POSTERIOR SHOULDER GIRDLE		
C7–T5	**4. Rhomboids**	Medial border of scapula
C1–C4	**5. Levator scapulae**	Upper medial border of scapula
Occiput/ligamentum nuchae/T1–12	**6. Trapezius**	Lateral clavicle/medial border of scapula
Sacrum/lower spine	**7. Latissimus dorsi**	Upper humerus
Lateral clavicle/acromion process/spine of scapula	**8. Deltoid**	Shaft of humerus
Fossa above scapular spine	**9. Supraspinatus**	Head of humerus (back)
Fossa below spine of scapula	**10. Infraspinatus**	Head of humerus (back)
Outer border of scapula	**11. Teres minor**	Head of humerus (back)
Anterior surface of scapula	**12. Subscapularis**	Head of humerus (front)

The Shoulder Girdle

The *shoulder girdle* is a yoke-like arrangement of bones suspended above the upper rib cage that provides a mobile structure for supporting the arms, allowing them a wide range of movement (Fig. 9-1). It consists of four bones: the two clavicles and the two scapulae, which "float" on top of the rib cage and are not attached to the spine or ribs except where the clavicles join the sternum.

Like the pelvis and legs, the shoulders and arms originally evolved as limbs for supporting and moving the body and are therefore similar in structure. Both the arm and the leg are supported by a girdle, or yoke-like framework, which provides a structure for a ball-and-socket joint for the limb. Both are composed of one long bone connecting to two bones, which form a lever system for moving the hand and foot. The hand and foot are quite similar in structure.

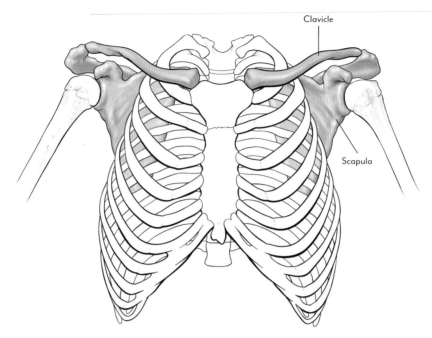

Fig. 9-1. The shoulder girdle acts as a yoke from which the arms are suspended.

In the upright human the arms hang freely and have become modified to be mainly manipulative. The shoulder girdle is attached to the axial skeleton only at the sternum (unlike the pelvic girdle, which is directly and firmly attached to the spine). This gives the scapulae a great deal of freedom and imparts to the arm a much greater range of movement than the pelvis does the leg.

The shoulder girdle consists of two bones: the clavicle and scapula (Fig. 9-2). The *clavicle* is a lever arm that mechanically joins the scapula and constrains the movements of the scapula. The *scapula* provides the socket for the humerus, is suspended from above, and moves independently of the ribs, providing a highly mobile support system for the upper limb. At the level of the shoulders, the rib cage appears to be very broad. But the arms are not attached directly to the rib cage, which is in fact very narrow at this point. It is the shoulder girdle and its complex musculature that gives breadth to the trunk. Yet although the clavicle and scapula form a cross-piece for supporting the arms, the scapula is part of the back and gains most of its support from the complex network of extensor muscles that make up part of the back.

Because the shoulder girdle functions as a system of muscles and levers, methods for treating the shoulder almost always address problems by stretching, strengthening, and releasing specific muscles. But the shoulder is in fact part of the larger postural system and can only be understood properly in this context.

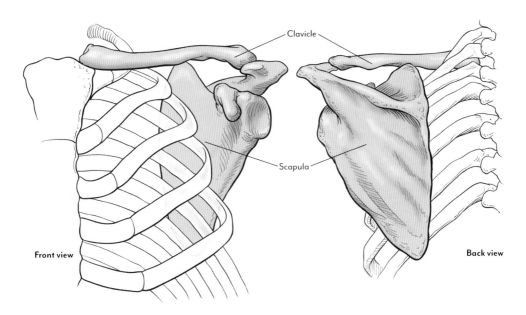

Fig. 9-2. The clavicle and scapula.

The Naturally Widened Shoulders

If you observe a young child of, say, two years old, it is easy to see that the shoulders are working as nature intended. There is an absence of tension, and the shoulders are integrated with the back in such a way that they are broad, relaxed, and strong. This expansive quality is not just a sign of youth; it is essential to how the shoulder works. When functioning properly, the shoulders are part of the back and completely free of tension. There is no shortening or narrowing in the pectoral region, because a two-year-old has not had the time to overwork the flexors, nor engaged in the kinds of activities that encourage this kind of shortening (Fig. 9-3).

Fig. 9-3. This two-year-old's upper arms remain in the plane of her torso and are not pulled forward in front of her trunk as she plays.

Many of us, however, are constantly performing activities that engage the flexor muscles of the hand, forearm, and shoulders, with the result that we narrow and shorten across the front of the shoulders. Coupled with the tendency to slump while sitting (which further contributes to the shortening and collapse of the shoulders), the flexors of the shoulders and arms in most adults become habitually shortened and the shoulders become narrowed in front.

THE AMAZING HUMAN SHOULDER

In humans, the scapula is a highly movable structure that adds considerably to the range of motion of the arm. If the scapulae were a simple, yoke-like crosspiece for the shoulder joints with sockets in each end, the range of movement of the arm would be quite limited (Fig. 9-4a). In order for the arm to be truly mobile, the socket itself—in other words, the scapula—needs to be movable (Fig. 9-4b and c). The articulations that make this possible are the sternoclavicular and acromioclavicular joints (Figs. 9-5 and 9-6).

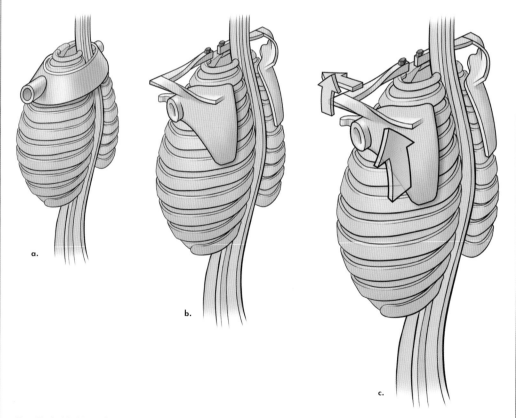

Fig. 9-4. *Unlike a fixed yoke, the shoulder girdle, which is not joined in back, can move in three planes.*

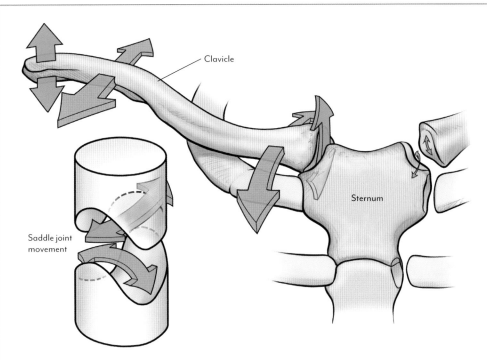

Fig. 9-5. The sternoclavicular joint is a saddle joint that allows the clavicle to move freely relative to the sternum.

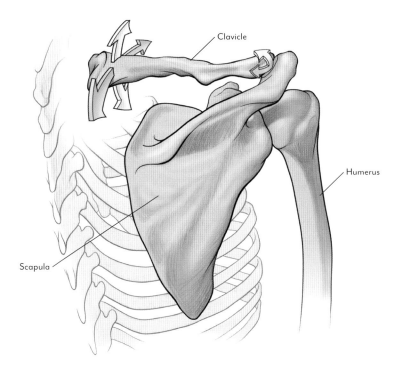

Fig. 9-6. The acromioclavicular joint allows the scapula to orient the shoulder joint in various directions.

THE ROTATOR CUFF

The shoulder joint, formed by the articulation of the head of the humerus with the glenoid cavity of the scapula, is the most mobile joint in the body (Fig. 9-7).

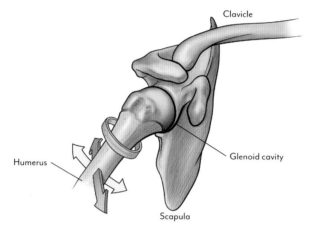

Fig. 9-7. The shoulder joint and its range of motion.

The humerus, which articulates with the shallow socket of the scapula, is held in place and moved by the rotator cuff, a set of four muscles that attach to the head of the humerus, forming a cuff around the joint that gives this group of muscles its name (Fig. 9-8).

The rotator cuff muscles are like two hands grasping the scapula that cup the ball of the humerus in the heel of the hand (Fig. 9-9).

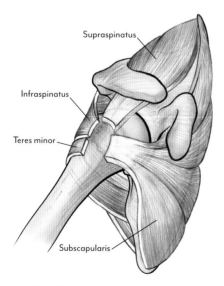

Fig. 9-8. The rotator cuff.

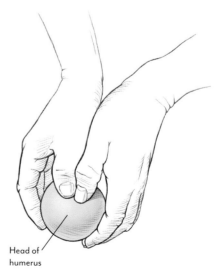

Fig. 9-9. The rotator cuff is like two hands cupping the head of the humerus.

In Fig. 9-10, the rotator cuff muscles and the muscles attaching to the medial border of the scapula (the rhomboids and serratus anterior) are shown in more detail.

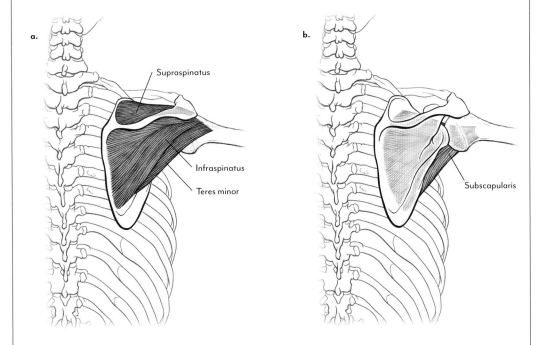

Posterior view of the right shoulder

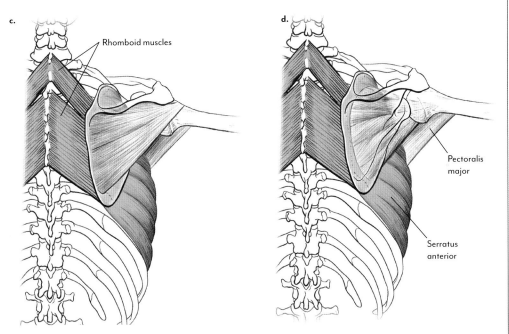

Fig. 9-10. a. & b. The rotator cuff consists of four scapular muscles attaching to different points on the head of the humerus; c. & d. The rhomboid and serratus anterior muscles attach to the medial border of the scapula.

SUBSCAPULARIS AND THE RAISED SHOULDER

When we use our shoulders in such a way that we narrow across the front of the chest and keep our arms raised for long periods, the rotator cuff muscles become over-worked and the joint loses mobility. When this narrowing is reduced, the rotator cuff muscles release; when this happens, it feels as if the shoulder joint softens, melts into the back, and regains flexibility and mobility. The rotator cuff muscles are thus essential to the proper function and mobility of the shoulder girdle.

Each of the rotator cuff muscles attaches to a different part of the head of the humerus. Infraspinatus and teres minor attach laterally and supraspinatus attaches at the top of the humerus. Subscapularis attaches to the front of the humerus and, when it shortens, raises and fixes the head of the humerus. When it releases, the joint softens and melts into the back (Fig. 9-11).

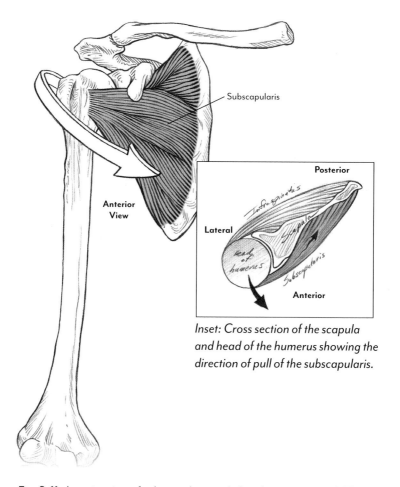

Inset: Cross section of the scapula and head of the humerus showing the direction of pull of the subscapularis.

Fig. 9-11. Anterior view of subscapularis with the rib cage removed. The subscapularis adducts the humerus and rotates it forward.

PECTORALIS MINOR AND
NARROWING THE SHOULDERS

One of the key functions of the pectoral muscles is to stabilize the scapula, as when we do a push-up. When these muscles become chronically shortened the scapula is pulled down and forward, narrowing the shoulders and pulling them out of the back. In order for the shoulder girdle to work properly, the shoulders must release in the pectoral region, where they tend to become narrowed and where the scapulae become pinned down to the ribs at one of their key attachment points: the coracoid process (Fig. 9-12).

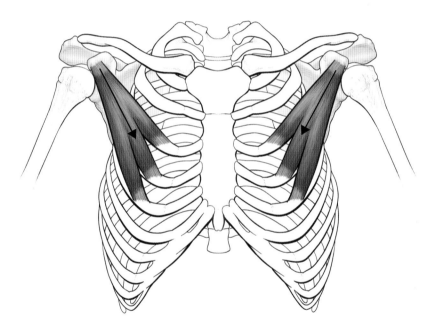

Fig. 9-12. When pectoralis minor shortens, the scapula is pulled down and forward (arrows), narrowing the shoulders.

PECTORALIS MAJOR

Pectoralis major is another key muscle that tends to become shortened, narrowing the shoulders. When these flexors release, the shoulders widen and, at the same time, link into the back. This allows the back muscles to work in a fuller and more supportive way. It also allows the ribs to move more freely, thus restoring deeper breathing. The shoulders are thus a crucial element in restoring the widened support of the back and the freeing of the ribs (Fig. 9-13).

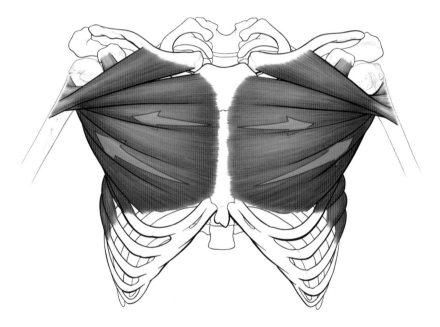

Fig. 9-13. Release of pectoralis major (arrows) enables the muscle to lengthen and the shoulders and ribs to assume their natural function.

THE CORACOID PROCESS AND THE SCAPULA

We've seen that the scapulae, which support the arms, are located mainly in back, where they are embedded in a network of muscles. On the front of the shoulder girdle, the most obvious muscle is pectoralis major, which covers much of the chest and inserts into the upper arms. But there is one other key muscle on the front that is critical, and it attaches not to the humerus but to the coracoid process: the bony projection on the front of the shoulder girdle that also serves as an attachment for the biceps and coracobrachialis muscles. When we habitually contract the pectoralis muscle and narrow the shoulders, it is useful to identify this landmark, which is so crucial to the proper working of the shoulder.

Further examination of the coracoid process clarifies why this attachment is so significant. If you look closely at the coracoid process, you can see that, although it terminates at the front of the shoulder, it actually projects from the scapula, which is located not in front but in back (Fig. 9-14). If the scapula is situated on the back of the rib cage, how is it possible for it to somehow reach forward to the front of the rib cage?

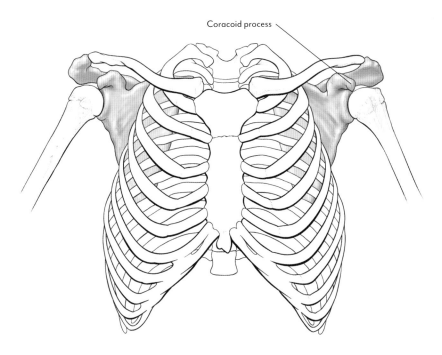

Coracoid process

Fig. 9-14. The coracoid process.

The answer is that the ribs are quite narrow at this point, and the scapula goes out beyond them to the sides and sends this projection forward. This is something you can mimic with your hand. Stand facing a friend and, wrapping your arms around them, place the palms of your hands on their back, about where their shoulder blades are. Now raise and abduct your thumbs so that they are facing directly toward you. The scapulae are represented by your hands and the coracoid processes by your thumbs; in this position, you can see that the scapula can rest in back but that the thumbs, pointing forward, provide an attachment for the pectoral muscles on the front of the chest (Fig. 9-15).

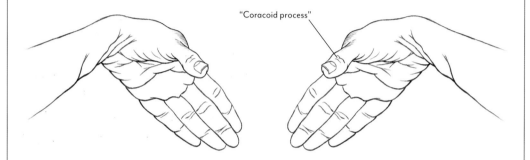

Fig. 9-15. Here the thumb represents the coracoid process, and the palm and fingers the scapula.

You can also see from this demonstration that, although the coracoid processes point forward, they are in fact part of the scapula, and that the pectoralis minor muscle, which attaches to this projection, is therefore a scapula muscle. There is a very good reason for this. The scapulae are highly mobile, and it is critically important to be able to anchor them from various directions. We can see how the muscles in the back do this; they anchor the scapula sideways, upward, and downward. But we also need to be able to anchor the scapula in front, which is where the coracoid process comes in. Because it projects forward, it becomes possible to act upon the scapula not just in back but in front, making it possible to anchor and move the scapula in virtually every direction. So although pectoralis minor is a chest muscle and attaches to the ribs in front, it is actually a mover and stabilizer of the scapula, and the coracoid process, projecting forward from the scapula like a thumb, is the key attachment upon which the pectoralis minor muscle can act to make this possible (Fig. 9-16).

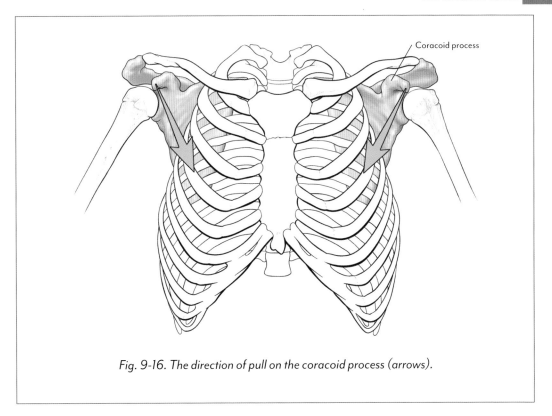

Coracoid process

Fig. 9-16. The direction of pull on the coracoid process (arrows).

SERRATUS ANTERIOR: THE HIDDEN FLEXOR

As a highly mobile support for the levers of the arm, the scapula requires the support of a large network of muscles that both move and stabilize it. If, as we've seen, the flexors on the front of the rib cage are overworked and shortened, the scapulae are pulled forward, which narrows the shoulders and prevents them from working properly. But there is another, less obvious muscle that attaches to the scapula and interferes with the shoulder girdle: serratus anterior, which is located not in front but in back.

Originating broadly on ribs 1–8, serratus anterior runs backward and underneath the scapula, attaching ultimately to its medial border (Fig. 9-17). Because this muscle is situated at the side of the chest and runs underneath the scapula, its action and purpose can be somewhat confusing. For instance, two other serrated muscles, serratus posterior and inferior, are both located on the back; because they attach to and act upon the ribs, they are clearly breathing muscles. With its broad attachment on the ribs, serratus anterior has a similar form and seems to be related to breathing.

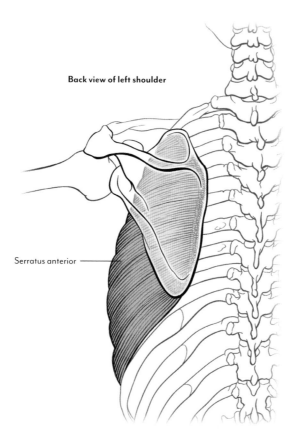

Back view of left shoulder

Serratus anterior

Fig. 9-17. Serratus anterior.

But serratus anterior is not a rib muscle; its function is to move and stabilize the scapula, to which it is attached and upon which it acts on quite powerfully. How it does this becomes clearer when we look at the rhomboid muscles, which attach to the same region—the medial border of the scapula. The rhomboid muscle, as we can clearly see in Fig. 9-18, originates at the spine and runs laterally to attach to the scapula. Its function is quite clear: it pulls the scapula sideways and anchors it backward, toward the spine. When we consider that serratus anterior attaches to the same region but from the opposite direction, we can see that it acts antagonistically to the rhomboids and thus pulls the scapula away from the spine and forward, as when we punch with the arm or do a push-up and need to move or stabilize the scapula. Serratus anterior is designed to do this as a matter of course, but if it is chronically contracted, as it often is, it pins the scapula to the ribs, preventing the shoulders from naturally widening and interfering with the shoulder girdle as a whole (Fig. 9-19). Serratus anterior thus acts as a hidden flexor that interferes with the shoulder girdle; when it releases, the scapulae go back and the shoulders become wonderfully widened and free.

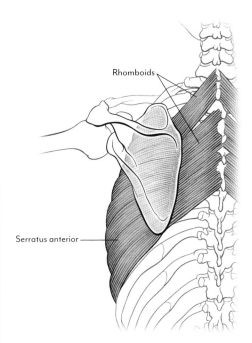
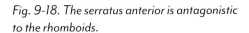
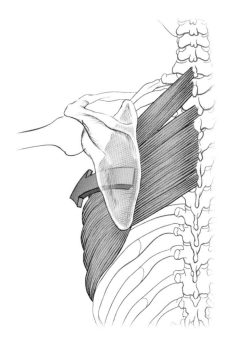

Fig. 9-18. The serratus anterior is antagonistic to the rhomboids.

Fig. 9-19. When it is chronically shortened, serratus anterior stretches the rhomboids and pulls on and fixes the scapula, preventing the shoulder girdle from widening.

MUSCLES ATTACHING TO THE UPPER ARM

Two key muscles attach to the upper humerus, pectoralis major in front and latissimus dorsi in back (Fig. 9-20). These muscles are designed to powerfully anchor and move the arm, but if they are chronically contracted, they pin the shoulders down and forward. For the shoulders to work properly, the arms must widen apart where these muscles attach into the upper arm, allowing these muscles to release and the shoulders to widen apart. This eases the downward pressure of the shoulder girdle and arms on the rib cage, thereby allowing the trunk and ribs to open and release in front. This also allows the back muscles to widen and fill out so that the scapula can become fully supported within the network of muscles, and the rotator cuff muscles around the shoulder joint can release and allow the scapulae to reintegrate into the back.

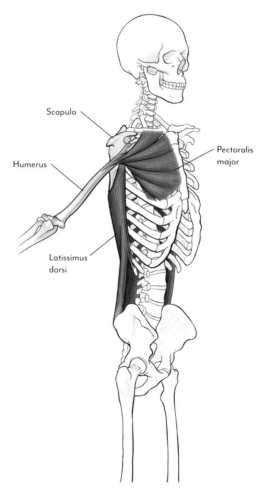

Fig. 9-20. The pectoralis major and the latissimus muscle attach into the upper humerus.

THE TRAPEZIUS AND DELTOID MUSCLES

The trapezius connects into the clavicle and shoulder blade and continues into the deltoid muscle, forming a connection between the support structure of the shoulder girdle and the muscles surrounding and supporting the shoulder joint (Fig. 9-21).

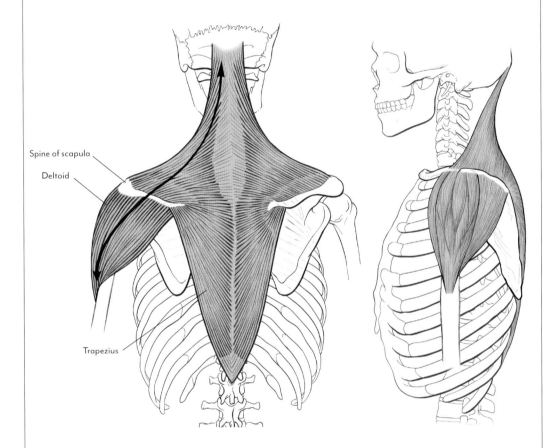

Spine of scapula

Deltoid

Trapezius

Fig. 9-21. The arrow indicates a continuous connection of the trapezius and deltoid muscles from the occiput to the humerus.

SUSPENSORY MUSCLES OF THE SHOULDER GIRDLE

The shoulder girdle is suspended by the trapezius and levator scapulae muscles in back and the sternocleidomastoid muscle in front. This enables it to float above the rib cage so that it can be moved freely and can function independently of the ribs (Fig. 9-22).

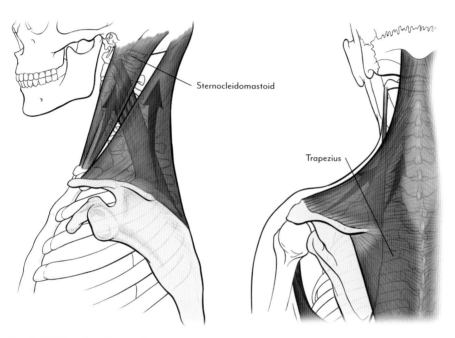

Sternocleidomastoid

Trapezius

Fig. 9-22. The shoulder girdle is suspended from above by the trapezius and sternocleidomastoid muscles (see arrows).

The levator scapulae is a crucial part of the network of muscles supporting the scapula and is continuous with supraspinatus (Fig. 9-23). The levator scapulae originates from the transverse processes of the first four cervical vertebrae and, angling downward, attaches to the apex of scapula. As its name suggests, it elevates the scapula, but its main job is to support the scapula and to assist in actions of the shoulder and arm. If it becomes overworked and stretched, the shoulders become raised and "held." To function properly, the shoulders must not only widen in front but release outward and forward so that they are neither collapsed nor raised, as we saw in the illustration of the little girl in Fig. 9-3.

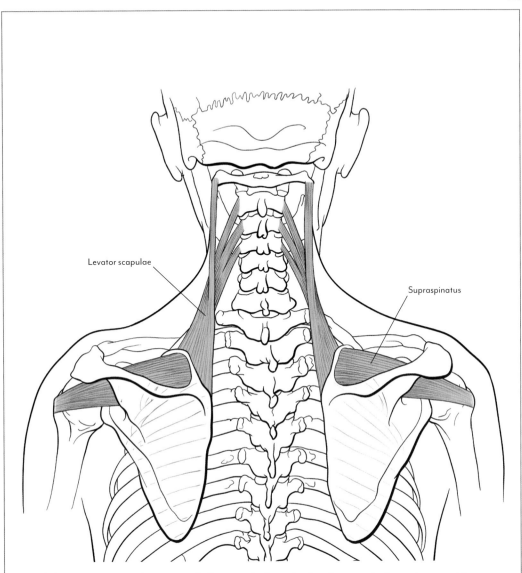

Fig. 9-23. Levator scapulae suspends the scapula from the first four cervical vertebrae and forms a continuous connection with the supraspinatus muscle, which attaches to the head of the humerus.

We've already considered upright support in terms of tensegrity principles. Here we see that the same tensegrity principles apply even to the larger moving muscles of the shoulder girdle, which is suspended from the head and the spine.

EXERCISE:
Widening the Shoulders and Using the Hands Like Feet

When we use our hands to do things, we must employ the flexor muscles of the arms and hands, which enable us to grasp and manipulate objects. But this is not the most basic function of the arms, which served our four-footed ancestors as forelimbs that supported the trunk above the ground by extending away from the trunk with the hands or paws opening onto the ground. Although we no longer use our arms primarily in this way, this extensor function is still essential to the proper use of the arms, which are meant to be supported by the extensor muscles of the back and shoulders even when we are grasping and manipulating objects. When working properly, the shoulders and back continue to widen even when we flex the arms or grasp with the fingers. This balanced use of the shoulders and arms can easily be seen in young children, whose shoulders and backs are broadly expansive even when they are grasping and manipulating objects.

In adults, however, the flexor function tends to dominate and interfere with this extensor function, causing us to habitually tighten the back muscles, narrow across the shoulders, and overuse the flexors of the arms when using our hands and arms. To restore proper support, we have to reengage the extensor function of the arms and back by putting ourselves in a position of mechanical advantage (such as sitting or monkey position) and by placing our palms on a flat surface with the arms working in a supportive way—that is, by using our hands like feet. When the arms are used in this way, we begin to elicit the supportive function of the back and arm muscles. Because the arms are now being used in a supportive way, we can stop the narrowing across the front of the shoulders and in the biceps, which allows the shoulders to spread apart and the back to work more fully.

Exercise 1: Placing Hands on Table Like Feet: The Procedure

To perform this procedure successfully, you should first be familiar with the inclined monkey position discussed at the end of Chapter 5. Because this procedure requires placing the palms face down on a table, it is important to make sure the table you are using is high enough. If you are fairly tall, you will have to straighten your arms to reach it. If this is the case, place a pile of books on the table for each hand, raising the table height so that you don't have to extend your arms and they can remain somewhat flexed at the elbow.

The exercise begins with going into the monkey position (see page 107), as follows:

1. Standing in front of the table with your feet slightly apart, give your directions so that you are allowing the head to nod forward, your back to lengthen, your gluteal muscles to let go, and your knees to go forward and away (Fig. 9-24a).

2. Take yourself into monkey and carefully go through your directions, allowing your head to go away from your pelvis and your pelvis from your head to lengthen your back, and your knees to go away from your pelvis and feet so that you are lengthening along the thighs and calves (Fig. 9-24b).

3. Place your hands on the table, palms facing down. Then release your neck again, and think of your head and pelvis going away from each other to lengthen the back. Lengthen up the front of the body from the hip bone to the mastoid process (Fig. 9-24c).

4. Come back to standing.

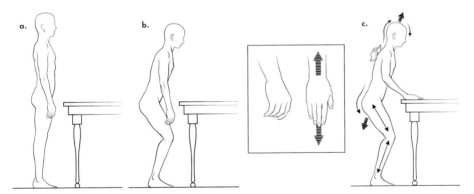

Fig. 9-24. Exercise: Hands on table like feet to widen the shoulders.

When you have become thoroughly familiar with this process, you can add several more directions for the hands and shoulders, as follows:

1. Again standing in front of the table, take yourself into monkey and carefully go through your directions (as shown in Fig. 9-24b).

2. Before raising your hands and placing them on the table, think of lengthening your fingers to point to the ground (inset). With your fingers leading, raise your hands at the elbow and place your hands on the table (Fig. 9-24c).

3. Free your neck to allow your head to go forward and up, to lengthen along the length of your spine, and to lengthen from the hip bone to the point on your skull just beneath your ear. Ask for length between the front of the hip joint and the back of the knee, and from the heels to the knees.

4. Then ask for length along the forearm from your wrist to your elbow, as if someone is pulling your elbows gently from behind. Do not stick your elbows out or tighten your shoulders to pull them up; allow the elbows to remain more or less dropped toward the sides of your body.

5. Finally, think of the inside of your upper arms so that the upper right arm is going away from the upper left and the upper left from the upper right. The two arms are spreading apart from each other to allow the widening of the upper part of the arm, which is where the pectoralis major muscle on the front of the body and the latissimus muscle in back attach.

Exercise 2: Directing the Arms in an Armchair

For this exercise, find a comfortable armchair and, following the illustration, give directions as follows:

1. Sit in the armchair with your arms fully supported on the arms of the chair.

2. Beginning with your left arm, think of allowing your fingers (a) to lengthen from your wrist (b). When you think of lengthening your fingers, don't forget to include your thumb. Pay attention to your fingers and hand for a minute or so, until you become kinesthetically aware of the length of your fingers and hand between these two points.

3. Ask for length in your forearm muscles from your wrist to your funny bone on the inside or lower part of your elbow (c), between point b and point c. Think about this until you are aware of the length of the forearm between your wrist and your elbow.

4. Now link up these two segments (a–b and b–c) so that you're aware of the length of your arm from fingertips to elbow. Think of your fingers lengthening away from your elbow, and your elbow lengthening away from your fingers.

5. Think of allowing the biceps region of your upper arm to release from your elbow to your shoulder joint (d), between point c and point d. Keep thinking this until you are kinesthetically aware of the length of your upper arm from elbow to shoulder.

6. Now link up segments a–c with segment c–d until you can feel the length of your arms from fingertips to elbow to shoulder as a continuous whole.

7. Turn your attention now to your right arm, and repeat these instructions from step 2, with points a1, b1, c1, and d1 on your right-hand side.

8. Turn your attention now to your shoulders and ask for length, or release, from your left shoulder joint to your right shoulder joint (d–d1), so that your right upper arm is going away from your left upper arm, and your left upper arm from your right upper arm. Spend a minute or so on this, thinking of widening across the shoulders until you are kinesthetically aware of the breadth or distance between d and d1.

9. Now think of linking up your left arm from a to d, across the shoulders from d to d1, and your right arm from d1 to a1. You should now feel the lengthening and widening of your arms and shoulder girdle from the fingertips of the left arm to the left shoulder, across the chest, and down to the fingertips of your right hand, as a continuous whole.

10. Returning to your body as a whole, think of your head going out of the length of your trunk, and think of allowing your knees to go forward. Since your forearms are lengthening from your fingertips to your elbows, and your shoulders are also widening to your elbows, you can simply think of directing to your elbows to lengthen your arms and to widen your shoulders.

11. You can now put all the directions together so that your head is going out of your lengthening and widening back, your knees are going out of your back at the other end, and your elbows are going out to the sides with the arms lengthening and the shoulders widening.

After performing this exercise, you should feel an opening and release of the shoulder girdle and an increased kinesthetic awareness, brought about through the thinking process just described.

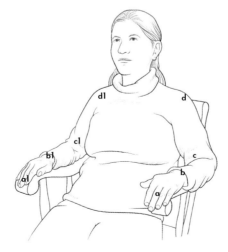

Fig. 9-25. Exercise: Directing the arms in an armchair.

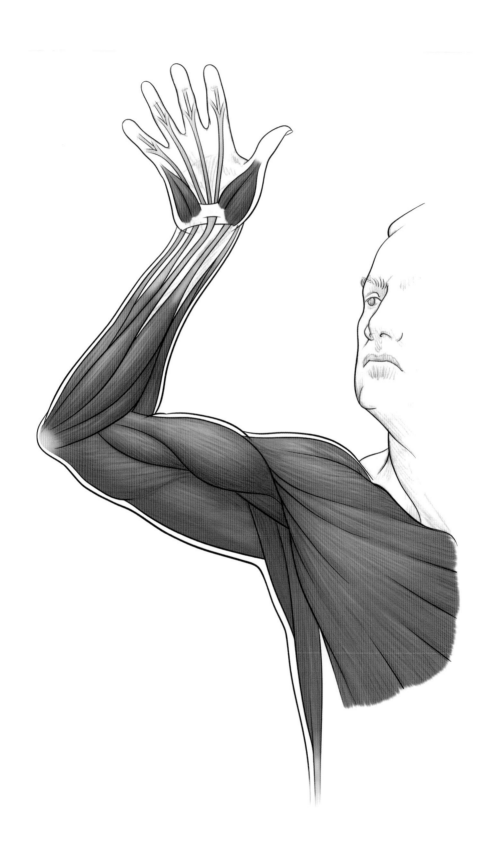

10

The Arms

The arm is a system of movable levers for positioning the hand, which is designed for grasping and manipulating objects. These levers move very freely at the shoulder girdle, which supports the arms and contributes to their range of motion. Unlike the legs, which have three muscular compartments, the arms are divided into only two—the extensors and flexors that form the spirals of the upper limb. In anatomical position, represented in the drawing on the next page, the extensors are on the back of the arm and the flexors are in front.

The Key to the Arms

When the shoulders widen, the arms release in the biceps, forearms, fingers, and thumbs. Thus, the key to the arms is the widening of the shoulders, as described in the previous chapter, coupled with the unimpeded use of the arm spirals letting go into length.

Contraindications

Many of us use the arms in a way that constantly engages the flexor muscles of the hand, forearm, and shoulders, causing a corresponding narrowing or shortening across the front of the shoulders, and shortening of the flexors of the arms and fingers.

WHEN THE ARM SPIRALS SHORTEN

1. The pectoral muscles shorten.
2. The biceps are shortened and overworked.
3. The forearm muscles are shortened and the hands and fingers become tight.
4. The flexors of the thumbs are shortened.

WHEN THE ARM SPIRALS LENGTHEN

1. The pectoral muscles release and the shoulders widen.
2. The biceps muscles release.
3. The flexors of the forearm lengthen and the hand opens up.
4. The thenar muscles of the thumbs release.
5. The extensors of the arms and the hypothenar muscles tone up.

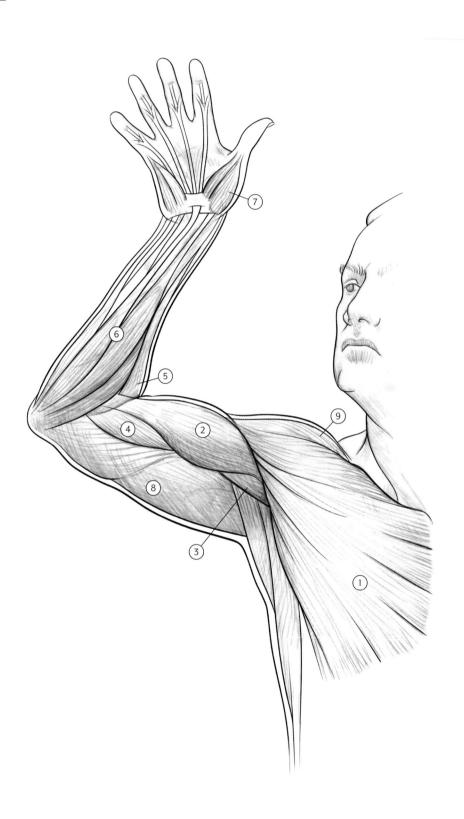

ORIGIN	MUSCLE	INSERTION
FLEXOR SPIRAL		
Coracoid process	**Pectoralis minor**	Ribs 3–5
Humerus	**1. Pectoralis major**	Sternum
Glenoid cavity/coracoid process	**2. Biceps brachii**	Radius
Coracoid process	**3. Coracobrachialis**	Shaft of humerus
Lower humerus	**4. Brachialis**	Ulna
Lower humerus	**5. Brachioradialis**	Distal radius
Common flexor origin	**6. Flexors of wrist and fingers**	Wrist and fingers
Wrist	**7. Flexors of thumb**	Thumb
EXTENSOR SPIRAL		
C7–T5	**Rhomboids**	Medial border of scapula
Glenoid cavity/humerus	**8. Triceps brachii**	Head of ulna
Lateral clavicle/acromion process/scapular spine	**9. Deltoid**	Shaft of humerus
Fossa above scapular spine	**Supraspinatus**	Head of humerus (back)
Fossa below scapular spine	**Infraspinatus**	Head of humerus (back)
Outer border of scapula	**Teres minor**	Head of humerus (back)
Anterior surface of scapula	**Subscapularis**	Head of humerus (front)

The Upper Limb

The arm is made up of two levers that hinge at the elbow (Fig. 10-1); these levers are quite long so that we can reach and position the hand in a wide arc around the body and in widely varying movements. The pattern of bones in both the arm and the leg are the same: one long bone, two bones, a cluster of wrist (or ankle) bones, followed by the metacarpals (hand) or metatarsals (foot), and terminating with the phalanges of the fingers and toes (see Fig. 12-1 for the lower limb).

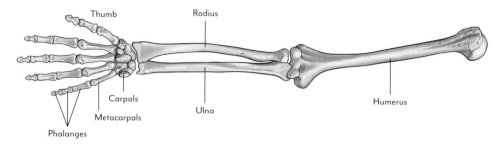

Fig. 10-1. Bones of the right arm, palmar view.

Although we retain the ability to support weight on our arms, the main function of the arm muscles is to move and position the hand for the purpose of grasping and manipulating objects. To use our arms efficiently, we must be able to engage the flexors without compromising the lengthened support of the shoulder girdle and arms. This makes it possible to use our hands, even for sustained periods of time, with a minimum of effort and interference.

As a complex lever system, the arm and hand are moved by two groups of muscles: the muscles on the dorsal or outer surface of the arm that extend the joints, and the ventral or inner muscles that flex the joints. Each of these muscle groups spirals down the arm; to use the shoulder girdle and arm efficiently, we must be able to engage the flexors without compromising the lengthened use of either of these groups of muscles.

The Spirals of the Arm

Let's look at the two spirals of the arms in greater detail, beginning with the back or extensor spiral, which includes the extensors of the arm, forearm, and hand.

The Extensor Spiral

The extensor spiral of the arm originates at the back of the shoulder girdle with the trapezius, rhomboids, and levator scapulae muscles, which begin at the midline of the spine and converge into the scapulae (see Chapter 9). Continuing from the scapula, the rotator cuff muscles—supraspinatus, infraspinatus, subscapularis, and teres minor—run in a similar direction to attach to the humerus, followed by the triceps brachii muscle, which crosses the back of the shoulder joint to attach into the olecranon, or elbow. The extensor line continues with the muscles on the dorsal surface of the forearm, which cross the wrist joint and attach to the wrist bones and fingers on the dorsal side of the hand (Fig. 10-2).

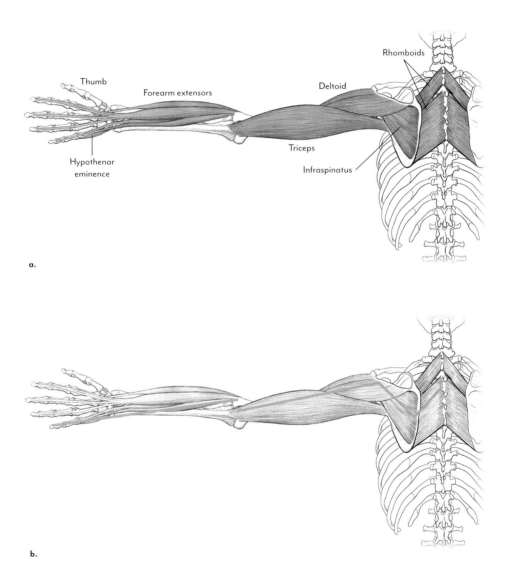

Fig. 10-2. Left arm, dorsal view: a. The extensor spiral muscles; b. Tracing the extensor muscles.

THE RHOMBOIDS AND INFRASPINATUS MUSCLES

The rhomboids originate at the spinous processes of the spine and attach to the medial border of the scapula—a line that continues with the infraspinatus muscle, which runs laterally to attach into the humerus (Fig. 10-3). When the shoulders are narrowed, these muscles shorten and lose tone. When the flexors at the front of the shoulders release, these muscles lengthen from the spine right out to the arm. Thus, the shoulder widens not only in front but in back, gaining its full support where it is needed, in the back.

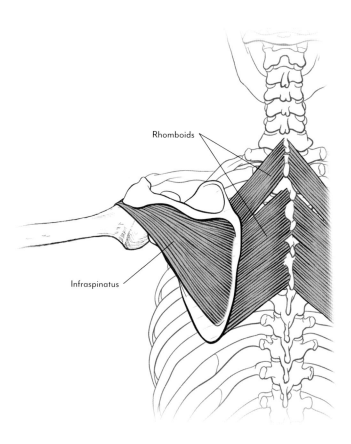

Rhomboids

Infraspinatus

Fig. 10-3. Rhomboids and infraspinatus shown with the arm abducted.

The Flexor Spiral

The flexor spiral of the arm originates at the front of the shoulder girdle with the pectoralis major and minor muscles. Pectoralis major originates at the sternum and attaches into the upper humerus; pectoralis minor originates at the third, fourth, and fifth ribs and inserts into the coracoid process of the scapula. The coracoid process is the key origin for the biceps brachii muscle and coracobrachialis, which insert into the arm and forearm. The flexor line continues into the forearm with a complex of muscles, mostly originating at the medial epicondyle of the humerus or funny bone, some tapering into tendons that attach to the wrist, others crossing the wrist and inserting into the fingers and, very importantly, into the thenar muscles of the thumb (Fig. 10-4).

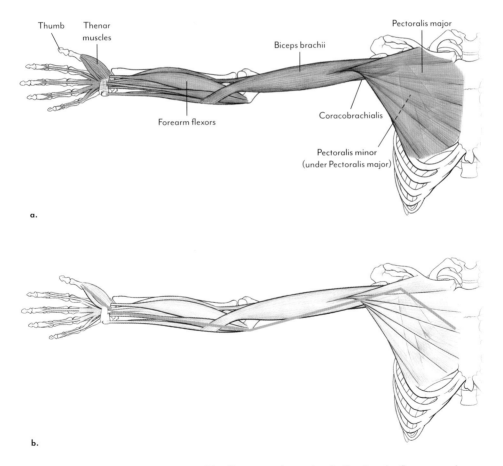

Fig. 10-4. Right arm, ventral view: a. The flexor spiral muscles; b. Tracing the flexor muscles.

THE VENTRAL AND DORSAL COMPARTMENTS OF THE ARM

The arm, unlike the leg, is divided into only two compartments—extensors and flexors—which are clearly delineated in both the upper and the lower arm. The flexors are linked into the pectoral region of the chest; the extensors are linked into the back (Fig. 10-5).

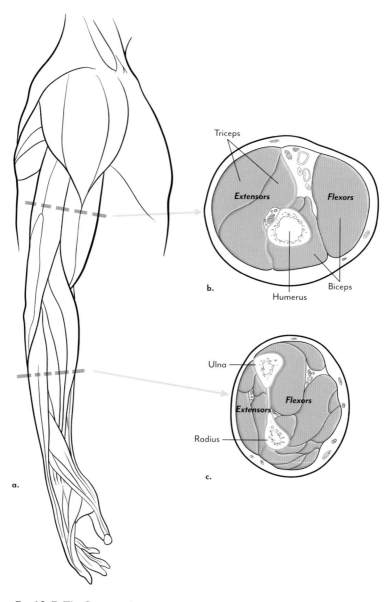

Fig. 10-5. The flexor and extensor compartments of the arm: a. Cross-sections; b. Upper arm—note the dominance of the extensor compartment; c. Lower arm—here the flexors, which work the fingers, predominate.

PECTORALIS MINOR AND BICEPS BRACHII

Pectoralis minor and biceps brachii share an attachment at the coracoid process. As a key flexor of the arm at the elbow, the biceps brachii muscle attaches to the radius and, as such, is also a supinator of the forearm. But it also attaches to the flexor fascia of the forearm. In this sense, it is clearly continuous with the flexion function of the hand, and establishes a line of communication right up the biceps and through to the pectorals. This line from the flexors of the hand to the armpit is important as it explains how overworking the fingers relates directly to narrowing of the shoulders (Fig. 10-6).

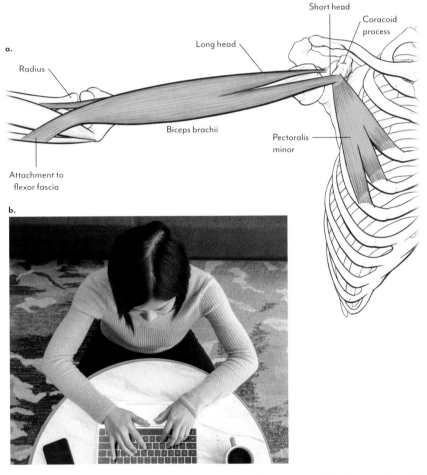

Fig. 10-6. a. Pectoralis minor and biceps brachii are key muscles of the flexor spiral of the arm; b. Notice how this girl is narrowing her shoulders, pulling them forward and down as she types.

THE THENAR AND HYPOTHENAR EMINENCES

The intrinsic muscles of the thumb—the muscles that move the thumb in relation to the hand—form the fleshy pad of the thumb called the *thenar eminence* (Fig. 10-7).

The intrinsic muscles of the little finger occupy the ulnar side of the hand and form the *hypothenar eminence*, the muscular pad on the side of the hand below the little finger (Fig. 10-8).

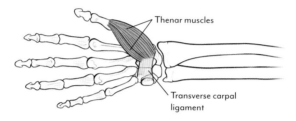

Fig. 10-7. Muscles of the thumb: the thenar eminence (right-hand palmar view).

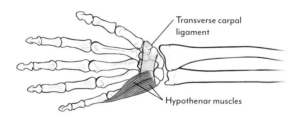

Fig. 10-8. Muscles of the little finger: the hypothenar eminence (right-hand palmar view).

Although the thumb seems to form the strong side of the hand and the pinky finger the weak side, the opposite is in fact the case. Hold a needle delicately between thumb and forefinger and it becomes immediately apparent that, although we can grasp quite firmly with the thumb, it is the fine motor part of the hand and is designed for delicate work. In contrast, the pinky side of the hand is strong and sturdy, as you can see if you pound the table with your fist or push someone away with your hand. As part of the flexor system of the arm, the thumb is linked into the pectoral muscles on the front of the shoulder, while the pinky side of the hand, which is connected with extension of the arm, is linked into the back.

EXERCISE:
Releasing the Flexors of the Arm

Lie in the semi-supine position. Extend and abduct your arms with your palms facing upward, being careful not to arch your back as you do this. With your palms up, your hands may not contact the floor so if you need to, place a pillow under them for support. In this position, which is ideal for releasing the flexors of the arms and pectoral region, you can trace a continuous line from pectoralis minor, biceps, and forearm right into the thumb (Fig. 10-9). After you have gone through the basic directions given in the exercise on page 59, add the following directions for the arms by thinking along the flexors on the upper or exposed side of each arm from breastbone to finger tips, following the flexor line shown in blue below. As you think of your arms releasing, do not forget to direct your head and knees. Keep the directions balanced and try not to over-focus on the arms to the exclusion of the other directions. As muscles start to undo, your arms will begin to lengthen.

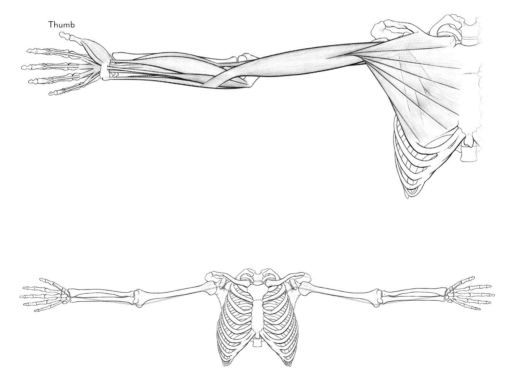

Thumb

Fig. 10-9. Exercise for releasing the flexors of the arm.

EXERCISE:
Releasing the Extensors of the Arm

Turn your attention now to the extensor muscles of your arms and shoulders. In the semi-supine position with your arms abducted and the pinky side of your hands in contact with the ground, the rhomboids and infraspinatus form a continuous line from the back into the triceps and dorsal muscles of the forearm to connect into the fingers and the hypothenar eminence (Fig. 10-10). Feel the back of your shoulders contacting the ground and allow your arm to lengthen out from the middle of your back, along your upper arms, through your elbow and lower arm and into your little finger. Notice the contact of the arm with the floor. As the muscles start to undo, your arms will start to lengthen and their contact with the floor will increase.

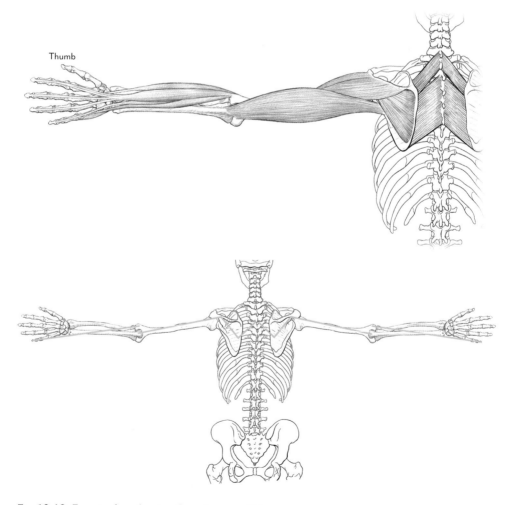

Fig. 10-10. Exercise for releasing the extensors of the arm.

EMBRYOLOGICAL ORIGINS OF THE SPIRALS OF THE UPPER LIMB

In Chapter 6, we saw that the spiral muscles of the trunk evolved in order to make twisting movements possible in land animals. In the early embryo, we could see that muscle segments formed on each side of the central axis and, after that, how they broke into flexors in front and extensors in back. To produce oblique lines of force, the flexors broke into three layers, two of which ran in oblique directions. These muscles were confined within the segments on each side of the body but, put together as a whole, they produced continuous spirals that wrapped around the entire trunk from pelvis to head.

Like the trunk spirals, the spiral patterns of the limbs appear in the embryo in clearly defined stages. As with other parts of the musculoskeletal system, the limbs develop in the early embryo from the mesoderm. During the fourth week of development, limb buds appear laterally and ventrally and grow into paddle-like structures (Fig. 10-11a). The limbs then begin to lengthen and extend forward, and the paddles begin to differentiate into recognizable hands (Fig. 10-11b).

It is during the eighth week that the limbs begin to reposition in ways that are more distinctly human, producing the spirals of the limbs (Fig. 10-12a). At this point, two things happen. First, the upper limbs rotate outward or laterally 90 degrees while, at the same time, the forearms remain pronated with the palms facing down. This creates

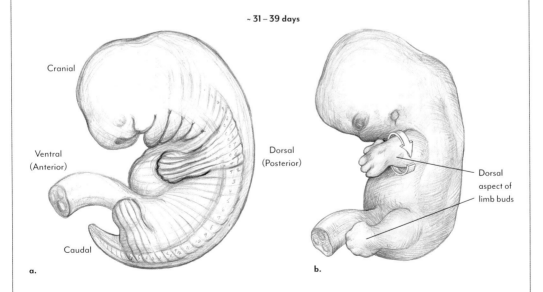

Fig. 10-11. a. Limb formation; b. Lateral rotation of shoulder.

a spiral in the arms, one part of which (the lower arm) is now rotated in relation to the upper part. The second thing that happens, a couple of days later, is that the lower limb rotates 90 degrees inward or medially so that the knees now point forward (Fig. 10-12b). Unlike the spirals of the arms, which are less distinct, the spirals in the legs are visible and dramatic. Before the rotation takes place, the extensors of the leg are more or less in back; after the rotation, the extensors are in front, demonstrating clearly that the muscles of the upper part of the leg have spiraled inward. The spiral of the arms is less obvious because, when the arms rotate outward, there is no dramatic or obvious spiraling action except for the fact that the lower arm remains pronated and thus, in relation to the outwardly rotated upper arm, seems to spiral inward.

The spiral of the upper arm becomes more apparent if you compare the dermatomes, or skin segments, of the lower and upper limbs. In Fig. 10-13, look first at the dermatomes of the leg and you can see how the segments clearly spiral around the thigh and down the leg, producing a spiral design like the old barbershop poles. Look now at the upper limb. At first glance, there do not appear to be any spirals: the flexors

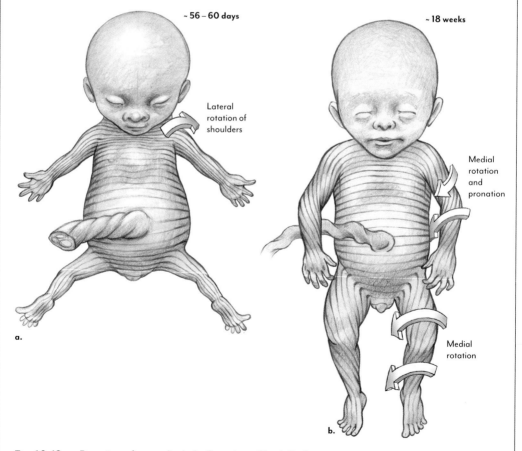

~ 56 – 60 days

Lateral
rotation of
shoulders

~ 18 weeks

Medial
rotation
and
pronation

Medial
rotation

a.

b.

Fig. 10-12. a. Rotation of upper limb; b. Rotation of both limbs.

of the trunk seem to extend directly to the flexors of the arm, and the extensors of the trunk to the extensors of the arm, with little spiraling of the muscles from front to back. This is because the arm is shown in anatomical position—that is, with forearms supinated and palms facing forward—and this is an artificial position that does not show the true position of the arms. In the embryo, the arms are pronated, not supinated; in this position, the forearm clearly spirals in relation to the upper arm. We can see this in the newborn infant as well, whose arms do not lie naturally in anatomical position but pronated with palms facing downward or backward (Fig. 10-14). When the arms are placed in this more natural position, the dermatomes do, in fact, wrap around the limb, particularly in the lower arm, clearly showing the spiral of the arm.

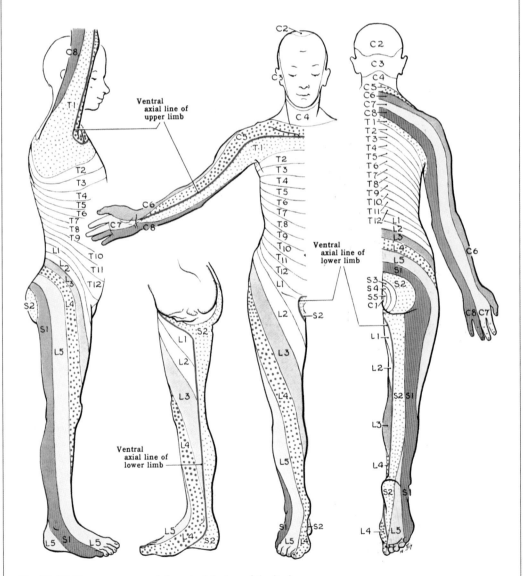

Fig. 10-13. Dermatomes and the spiral design of the limbs.

Fig. 10-14. Infant with naturally pronated forearms.

THE WRIST BONES AND THE HEEL OF THE HAND

When we support weight on our hands, what part of the hand do we use? Like the foot, the hand has a heel and, when we go on all fours, it is this part of the hand that bears most of the weight. Many infants go through a phase of crawling on all fours, reflecting that phase of our evolutionary heritage in which all four limbs were used for walking on the ground. Although we are all familiar with the hand's fine motor capabilities, we are much less familiar with this more primitive, yet crucial, use of the hand.

The heel of the hand is the area closest to the wrist joint and is made up of eight carpal bones (Fig. 10-15). When we think of the wrist, we are likely to think of the place where the wrist watch goes, just above the bumps on either side of the wrist. This is actually not your wrist but your forearm. More accurately, the wrist joint is where the hand articulates with your forearm. The wrist also refers to the wrist bones themselves—the carpal bones that form the base of the hand. Anatomy books refer to these as wrist bones (in contrast to the bones of the hand); but the wrist bones form the heel of the hand, so we could say that the wrist is actually part of the hand itself.

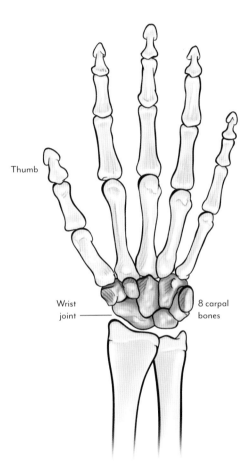

Fig. 10-15. Palmar view of left hand showing the carpal bones of the wrist.

Two areas that relate to the wrist are the pad of the thumb, which as we saw is called the *thenar eminence*, and the fleshy pad under the little finger that forms the side of the hand, called the *hypothenar eminence*. Both of these pads are fleshy because they are made up of flexor muscles that move the thumb and pinky, as well as fatty tissue that adds to the padding needed in these areas. The heel of the hand includes the lower part of each pad, as well as the gap in between, which is the carpal tunnel (Fig. 10-16).

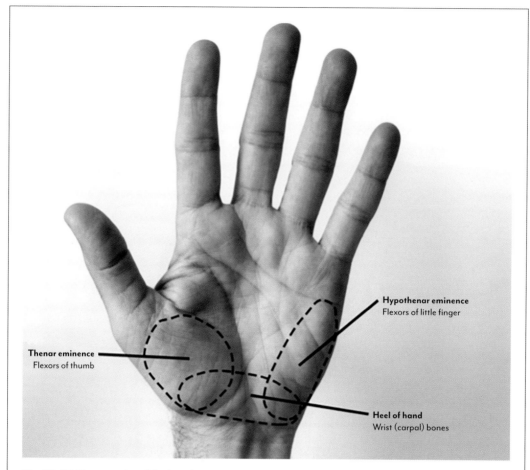

Hypothenar eminence
Flexors of little finger

Thenar eminence
Flexors of thumb

Heel of hand
Wrist (carpal) bones

Fig. 10-16. Topography of the hand.

Although the hands are designed for grasping and manipulating objects, they can also be used for bearing weight—as when we go down on all fours or push a large object with open palms—in which case they are functioning more like feet. When we use the hands in this more primitive way, the flexors of the forearm are less active and, with the arms extended, the palms are more open. To recover this primitive, open use of the hand, it is useful to spend time exploring how to place the heel of the hand on a surface in order to use the hands like feet. To experiment with this, return to "Exercise 1: Placing Hands on Table Like Feet" on page 206. After you have gotten to step 3, try releasing in your wrist joints to allow the heels of your hands to go onto the table surface—making sure that you are not tightening your fingers—before removing your hands and coming back to standing. Allowing your hand to open onto the surface encourages a new and more coordinated use of the shoulder and upper limb (see the exercise called "Extending Hand at Wrist to Form an Extensor Grip" later in this chapter). When we use the hand in this way, it has little connection with grasping and therefore enables the hand, arm, and shoulder to work in a completely new and coordinated way.

THE FOREARM MUSCLES AND
THE COMMON FLEXOR ORIGIN

Most of the flexors of the forearm originate at the medial epicondyle of the humerus—otherwise known as the "funny bone" (Fig. 10-17). Because we tighten and chronically shorten these muscles, they play an important role in the proper working of the arm.

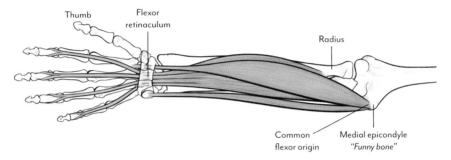

Fig. 10-17. Palmar view of right hand showing the forearm flexors and the common flexor origin.

THE FOREARM MUSCLES AND
THE COMMON EXTENSOR ORIGIN

Just as the flexors of the forearm have a common origin on the humerus, so do the extensors, most of which originate at the lateral epicondyle of the humerus (Fig. 10-18). If the flexor muscles become overworked and shorted, so do the extensors, and as we saw earlier in the chapter, both these muscle groups have a direct connection into the shoulders and can affect arm movement as a whole.

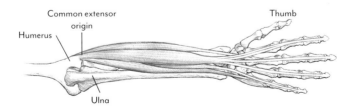

Fig. 10-18. Dorsal view of right hand showing the forearm extensors and the common extensor origin.

THUMB VERSUS FINGERS

Depending on how we orient the hand at the wrist, we can facilitate the lengthening and release of fingers and thumb. With the hand deviated at the wrist, the thumb comes out of the flexors in a straight line; this is conducive to releasing the thumb out of the flexors of the forearm and upper arm and, ultimately, out of the pectoral region of the shoulder (Fig. 10-19).

With the hand not deviated, the fingers come out of the flexors in a straight line. By directing the fingers in this position, you can facilitate lengthening and release of the flexors in the forearm (Fig. 10-20). Both of these exercises can be performed while sitting at a desk with your elbows resting on the table surface, or in the semi-supine position with your elbows on the ground with your fingers pointing upwards.

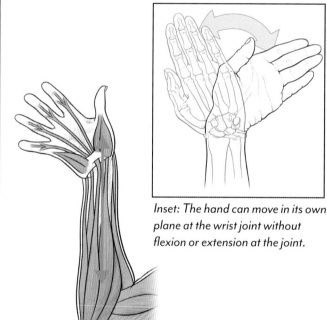

Inset: The hand can move in its own plane at the wrist joint without flexion or extension at the joint.

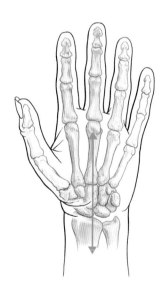

Fig. 10-19. With the hand deviated at the wrist, the thumb comes out of the flexors in a straight line.

Fig. 10-20. With the hand not deviated, the fingers come out of the flexors in a straight line.

THE TENDON COMPLEX OF THE FINGERS

The large movements of the hand and fingers are controlled by the flexors and extensors of the forearm, which taper into tendons that cross the wrist (Figs. 10-17 and 10-18). To perform strong grasping movements, as when carrying a heavy object or hanging by the arms, the flexors need to be much stronger than the extensors. This is why the flexor compartment of the arm, which we can see in the cross-section (Fig. 10-21a), is much bulkier and contains more muscles than the extensor compartment. The flexors also require more tendons, as we can see in the cross-section of the hand at the wrist (Fig. 10-21b). The fingers have two joints that can flex independently between the three phalanges, and so a single tendon needs to be assigned to each joint for the fingers to benefit from full flexion. To extend the fingers, only one tendon is required for each of the fingers, which do not bend backwards at the joint but simply straighten.

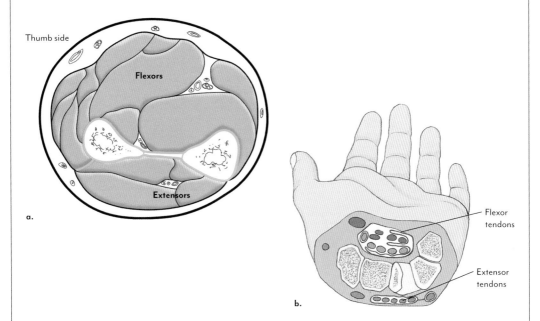

Fig. 10-21. a. Cross section of the left lower arm showing large flexor compartment and smaller extensor compartment; b. Cross section of the left wrist through the carpal bones showing the numerous flexor tendons and fewer extensor tendons.

EXERCISE:
Releasing the Thumb

1. Sit in front of a table and place your hands palms down on the table (Fig. 10-22a).

2. Leaving your thumbs and wrists on the table, extend your hands at the wrist so that your fingers are angled upward at about 45 degrees. The heels of your hands are now in contact with the table, and your thumbs lie more or less in a straight line with your forearms (Fig. 10-22b).

3. Direct your thumbs to lengthen away from your elbows.

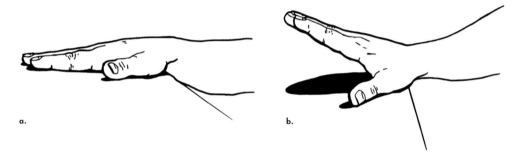

a.   b.

Fig. 10-22. Exercise for releasing the thumb.

Variation

This exercise can also be done while lying in the semi-supine position. With your hands resting on your midriff, ulnar deviate your hands at your wrists so that your fingers are pointing toward your feet. Leaving your thumbs and wrists on your belly, extend your hands at the wrist so that your fingers are pointing upward at 45 degrees, as in Fig. 10-22. Your thumbs, which are still in contact with your belly, now lie more or less in a direct line with your forearms. In this position, direct your thumbs to lengthen from your elbows. Try also raising your forearms so that, with your elbows still on the ground, your thumbs are pointing directly toward the ceiling; then direct your thumbs to lengthen and point upward, as in Fig. 10-19.

EXTENSOR GRIP OF THE HAND
AND THE LUMBRICAL MUSCLES

To hang by your arms or to forcefully grip an object, the muscles of the forearm—the extrinsic muscles of the hand—act powerfully on the hand and fingers. When grasping a scalpel or threading a needle, these muscles are less active; in this case, the intrinsic muscles of the hand, such as the thenar muscles that oppose the thumb to the fingers, are brought into play. A particularly interesting group of intrinsic muscles, the lumbricals, flexes the first digits of the fingers while simultaneously extending the second and third digits (Fig. 10-23). Just as the thenar muscles make it possible to oppose the thumb to the fingers, this muscle makes it possible to oppose the fingers to the thumb without involving the larger muscles of the forearm.

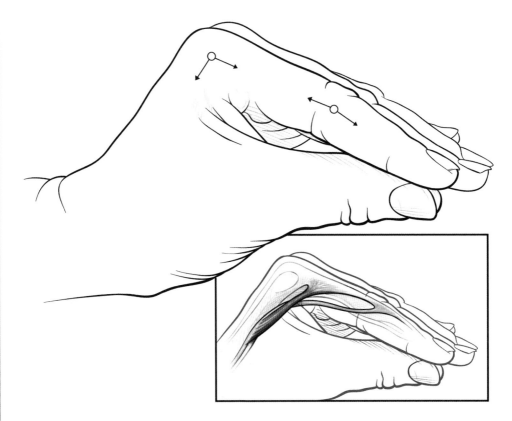

Fig. 10-23. The lumbrical muscles flex the fingers at the knuckles, keeping them extended.

EXERCISE:
Extending Hand at Wrist to Form an Extensor Grip

1. Sit in front of a table and place your hands palms down on the table
   (Fig. 10-24a).

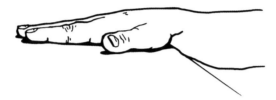

Fig. 10-24a.

2. Leaving your thumbs and wrists on the table, extend your hands at the wrist
   so that your fingers are angled upward at about 45 degrees. The heel of your
   hand is now in contact with the table, and your thumbs lie more or less in a
   straight line with your forearm (Fig. 10-24b).

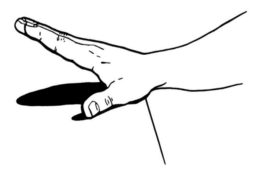

Fig. 10-24b.

3. Flex your fingers at the knuckles so that they point down to the table and
   remain extended from the knuckles and, as you do so, slide the thumb of each
   hand toward the fingers. The fingers, and in particular the index finger, now
   oppose the thumb (Fig. 10-24c).

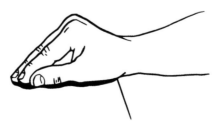

Fig. 10-24c.

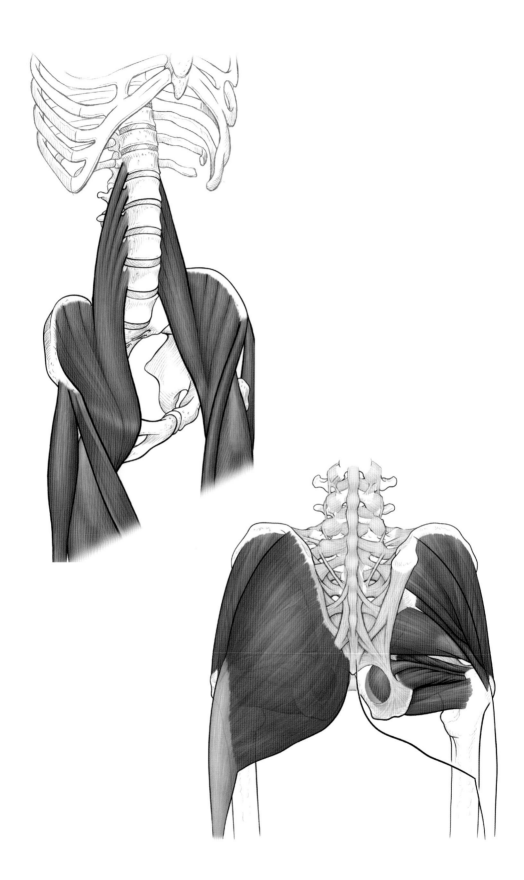

11

The Hips

Like the scapula of the shoulder, the pelvis provides bony sockets for the levers of the limbs. In contrast to the shoulders, however, which are highly mobile, the lower limbs are designed for support and propulsion on two legs. For this purpose, the pelvis is directly connected to the spine and acted upon by a series of muscles that extend, flex, rotate and, very importantly, maintain the stability of the hip joints and the trunk at the hip. This makes it possible to transmit forces from the legs directly to the trunk while maintaining the lengthened support of the head and trunk on two legs.

The Key to the Hips

When the gluteal muscles and deep muscles of the hip release across the back of the pelvis, this allows the legs to undo out of the hips. At the same time, the flexors that cross the front of the hip must release to allow the thighs to lengthen out of the pelvis. This allows the hip muscles to perform their postural duties and the hip joints to remain free, even while supporting weight and walking on two legs.

Contraindications

Many of us brace the legs and hips joint when we stand and walk. The habit of slumping further contributes to the tendency to lose tone in the hip muscles and, in response, to grip and tighten the hips in order to stand or to sit upright. This causes the hip joints to become habitually fixed and shortened.

WHEN WE SINK INTO AND STIFFEN THE HIP

1. The hip muscles lose tone.
2. The deep muscles of the hip become shortened and the hip joint becomes fixed.
3. The gluteal muscles grip and tighten.
4. The flexors of the hip shorten.

WHEN WE RELEASE THE HIP JOINT

1. The hip muscles tone and release and the hip joint regains mobility.
2. The gluteal muscles release.
3. The hip flexors release and the leg lengthens out of the hip.

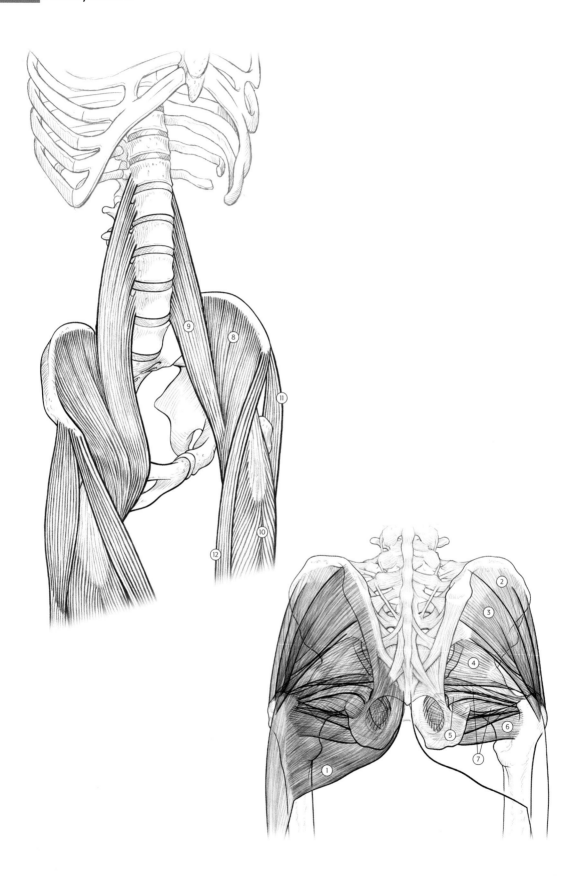

ORIGIN	MUSCLE	INSERTION
EXTENSORS		
Ilium/sacrum	**1. Gluteus maximus**	Iliotibial (IT) band/shaft of femur
Inner border of ilium	**2. Gluteus medius**	Greater trochanter
Surface of ilium	**3. Gluteus minimus**	Greater trochanter
Anterior sacrum/ Sacrotuberous ligament	**4. Piriformis**	Greater trochanter
Obturator membrane	**5. Obturators**	Greater trochanter
Ischium	**6. Quadratus femoris**	Below greater trochanter
Ischium	**7. Gemelli**	Tendon of obturator muscles
FLEXORS		
Anterior surface of ilium	**8. Iliacus**	Lesser trochanter
T12, L1–L4	**9. Psoas major**	Lesser trochanter
Ilium above acetabulum	**10. Rectus femoris**	Tibial tuberosity
Anterior iliac crest	**11. Tensor fascia latae**	IT band
Iliac crest	**12. Sartorius**	Medial epicondyle of tibia

The Pelvic Girdle

In our upright posture, the trunk is balanced vertically on the heads of the femurs, freeing the arms for manipulative purposes. But achieving upright posture is not simply a matter of raising the trunk on the rear legs. To be balanced on the femurs, the trunk requires a stable base of support, which is provided by the pelvis (Fig. 11-1a). In four-footed animals, both the fore and hind limbs bear the weight of the head and trunk. In humans, the pelvis not only bears weight; it bears the entire weight of the trunk, which now rests solely the on the legs—a tremendous increase in responsibility (Fig. 11-1b). For this purpose, the pelvic bones, which are strong and sturdy, become firmly joined to the sacrum to form the sacroiliac joints. In contrast to the shallow and mobile joints of the shoulder, the sockets for the femurs became deeper, providing very stable joints for the femurs (Fig. 11-1c), and the hip joints are supported by a complex of muscles that support and move the femur at the hip.

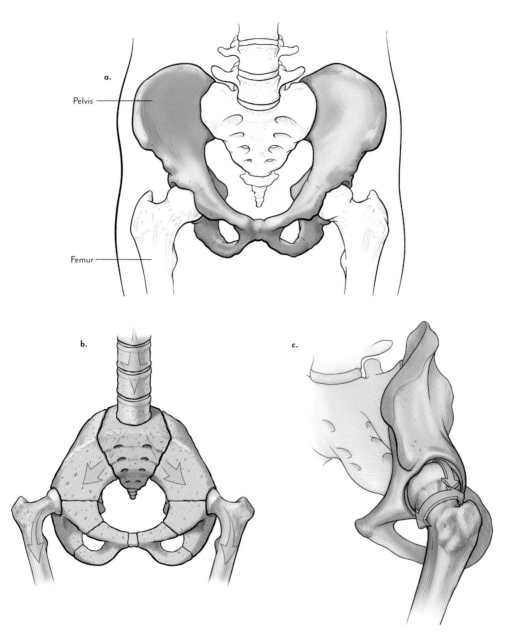

Pelvis

Femur

Fig. 11-1. a. With the iliac bones firmly attached to the sacrum, the pelvis forms a solid base for supporting the trunk on two legs; b. The arch design of the pelvis transfers the weight of the trunk into both legs; c. The hip is a ball-and-socket joint that provides both stability and mobility.

The Iliopsoas Complex

As we saw in Chapter 5, to make upright posture possible, the human spine had to develop a lumbar curve, which creates instability in the lower back. The iliopsoas muscle, which crosses the inside of the pelvis as it passes from the lumbar spine to the femur, plays a crucial role in stabilizing the pelvis in relation to the back (Fig. 11-2). In conjunction with the erector spinae muscles, it maintains the support and length of the lower back (Fig. 11-3).

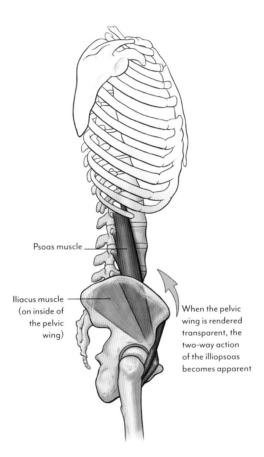

Psoas muscle

Iliacus muscle (on inside of the pelvic wing)

When the pelvic wing is rendered transparent, the two-way action of the illiopsoas becomes apparent

Fig. 11-2. The iliopsoas muscle is a key flexor of the leg at the hip. It also plays a crucial role in stabilizing the trunk and spine, helping to maintain the orientation of the pelvis and the balance of the trunk on the femurs.

When we slump and collapse the spine, however, the back muscles become contracted, and the iliopsoas muscle, which also becomes shortened, pulls on the lower back. In order to restore natural length and support in the trunk, the iliopsoas muscle must release across the front of the pelvis. Roughly corresponding to the freeing of the pectoral region of the chest, which allows the shoulder girdle to widen, this frontal release of the iliopsoas muscle allows the lower back to lengthen and the thighs to release out of the hips. It is thus a crucial element in restoring the lengthened support of the trunk (see Figs. 11-4 and 11-5). In this sense, the working of the lower back is completely intertwined with the pelvis and its connection with the trunk. To function in a coordinated manner, the legs must integrate with the lower back in such a way that they don't pull the pelvis and lower back out of balance or cause shortening in the iliopsoas muscles that support this region.

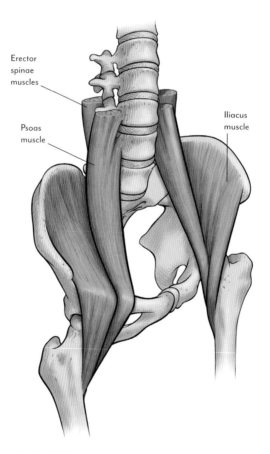

Erector spinae muscles

Psoas muscle

Iliacus muscle

Fig. 11-3. The iliopsoas and erector spinae muscles acting together maintain the lengthened support of the spine.

THE CONNECTION OF THE HIP FLEXORS TO THE BACK

Although the iliacus and psoas muscles are flexors, they form a direct connection to the spine and back, connecting the flexors in front of the hip to the extensors of the lower back (Fig. 11-4). When the psoas muscle is shortened and releases, it takes pressure off the lumbar spine. Where the iliacus muscle attaches to the pelvic rim, quadratus lumborum forms a connection in back to the ribs: when the iliacus muscle lets go and the knees go away, the quadratus lumborum muscle also lets go so that release of the hip flexors is part of restoring length in the lower back.

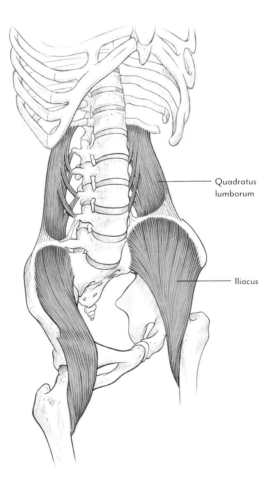

Quadratus
lumborum

Iliacus

Fig. 11-4. The iliacus and quadratus lumborum form a continuous line that connects the hip flexors with the extensors in back.

HIP FLEXORS

The iliopsoas complex is central to the lengthening action of the legs as extensions of the trunk (Fig. 11-5). If these muscles are contracted and shortened, the psoas muscle pulls on the lumbar spine and the two muscles shorten across the hip. When the legs are lengthening out of the trunk, the hip flexors can release across the front of the hip and pelvis. This release, which corresponds to the undoing of the pectoral muscles across the shoulder, is critical to restoring the lengthened support of the body as a whole.

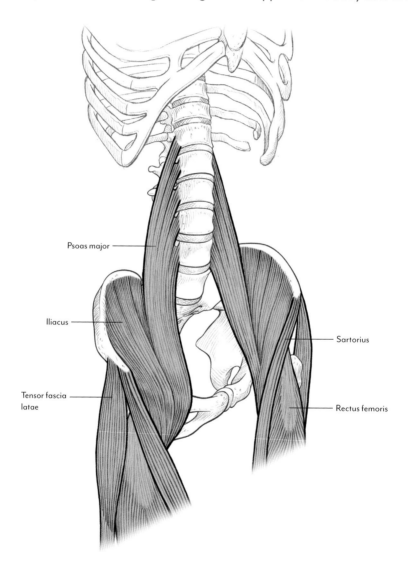

Fig. 11-5. The hip flexors are crucial to the lengthened support of the legs.

The Gluteal Muscles

The gluteal muscles are extensors of the hip and stabilizers of the leg. Gluteus maximus forms the bulkiest muscle of the buttocks region; it extends the trunk at the hip and assists in powerfully extending the leg. It consists of an upper and a lower section. The upper section attaches to the iliotibial (IT) band and the lower section attaches to the shaft of the femur (Fig. 11-6c). Gluteus minimus and gluteus medius are abductors of the hip and stabilize the hip when weight is placed on the leg in walking (Fig. 11-6a and b).

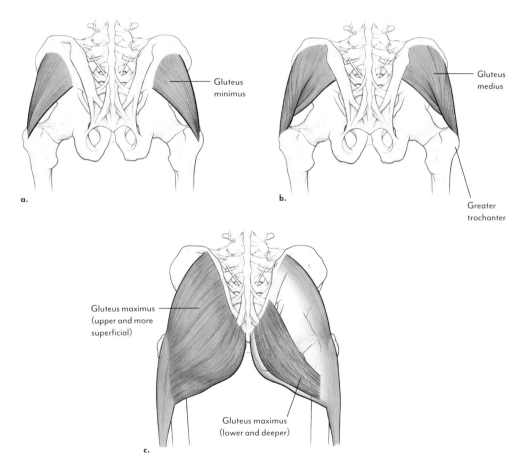

Fig. 11-6. The three gluteal muscles: a. Gluteus minimus; b. Gluteus medius; c. Gluteus maximus. Gluteus minimus and medius attach to the greater trochanter and stabilize the pelvis on the femur; gluteus maximus, whose insertions run into the IT band and femoral shaft, is a powerful extensor of the leg in upright posture.

TIGHTENING IN THE BUTTOCKS
AND RELEASING THE HIPS

We are designed to maintain upright support with the legs naturally lengthening out of the trunk. If we habitually slump and collapse, we disengage our upright support system and, when we stand, compensate by bracing and stiffening the hips and legs. The hip muscles become chronically clenched, laterally rotating the femur at the hip joint and fixing the joint—a tendency that becomes accentuated with postural twisting, which causes further clenching of the buttocks (Fig. 11-7a). (See also "Freeing the Hip Joint" later in this chapter.) When the legs lengthen and the hip muscles release, this takes pressure off the hip joint and the buttocks region unclenches, allowing the femur to rotate medially and restoring natural, effortless support (Fig. 11-7b).

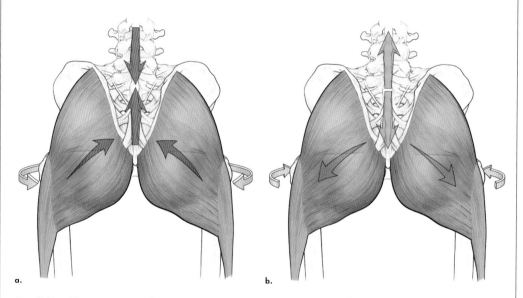

Fig. 11-7. a. To support ourselves against gravity, we overcompensate by gripping and shortening the gluteus maximus muscles. The result is over-rotation of the hip joints, tightening in the buttocks, and shortening in the lower back; b. When the gluteal muscles release, the buttocks unclench and the legs can lengthen out of a lengthening back.

Muscles of the Hip Joint: The Deep Six

There are six deep muscles of the hip: two obturators, two gemelli, quadratus femoris, and piriformis (Fig. 11-8). Originating at the pelvis and acting on the greater trochanter, these muscles are lateral rotators of the hip. The resulting action of all six muscles working together is to stabilize the head of the femur in its socket while assuring an outward rotational pull on the bone, which balances the inward pull of the iliopsoas muscle.

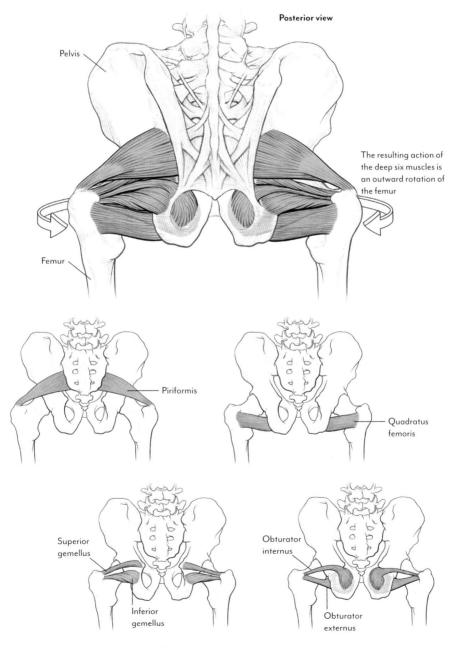

Posterior view

Pelvis

The resulting action of the deep six muscles is an outward rotation of the femur

Femur

Piriformis

Quadratus femoris

Superior gemellus

Inferior gemellus

Obturator internus

Obturator externus

Fig. 11-8. The deep muscles of the hip joint.

THE FAN MUSCLES OF THE HIP

All the muscles of the hip (with the exception of gluteus maximus, which does not actually attach to the hip) spread out from the greater trochanter to form a fan attaching to different points on the pelvis (Fig. 11-9). Their role can be likened to that of the rotator cuff muscles of the shoulder (page 192), which act on and support the shoulder joint. Unlike the freely moving arms, however, the legs are designed for support and locomotion on two feet. For this purpose, the hip muscles maintain postural stability and upright support even while allowing the legs to swing freely at the hip joints.

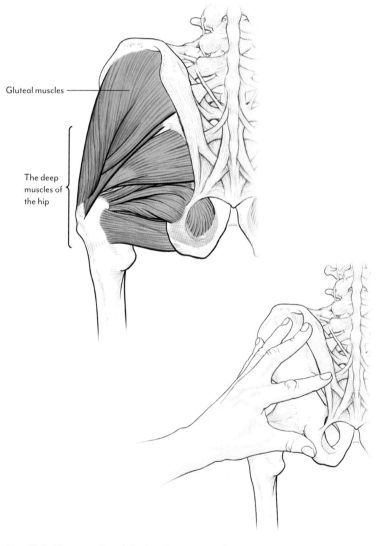

Gluteal muscles

The deep muscles of the hip

Fig. 11-9. The muscles of the hip fanning out from the greater trochanter resemble the fingers of the hand spreading out from the palm.

FREEING THE HIP JOINT

The deep muscles of the hip act as lateral rotators of the hip, but they become contracted and fix the joint when we are sitting, standing, and walking (Fig. 11-10). When these muscles let go, the hips rotate inwardly as part of the release of the outer or extensor leg spiral (Fig. 11-11).

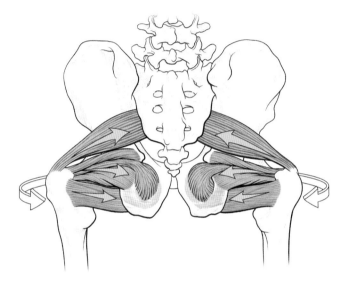

Fig. 11-10. When the deep muscles of the hip contract and shorten, the joint loses mobility and the head of the femur becomes fixed.

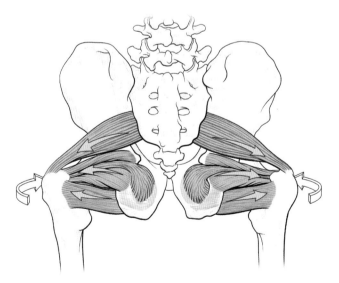

Fig. 11-11. In this case, when the muscles release, the joint can once again move freely.

PIRIFORMIS:
A KEY STABILIZER OF THE HIP AND SACRUM

One of the key muscles of the hip region is the piriformis muscle, so named because of its resemblance to a pear. The piriformis originates at the anterior surface of the sacrum and, passing laterally and slightly downward, inserts into the greater trochanter of the hip (Fig. 11-12a). As a hip muscle, its most obvious function, along with the other deep muscles in this region, is to laterally rotate the hip and to assist in extending the hip (Fig. 11-12b).

But the piriformis has another key function. The spine has twenty-four moveable vertebrae that terminate at the sacrum, which is nested between the iliac bones to form the sacroiliac joints. Five vertebrae are fused together to form the triangular-shaped sacrum; and the weight from the sacrum is transferred through the sacroiliac joints to the pelvis. When we shift our weight onto one leg or lean to one side, the

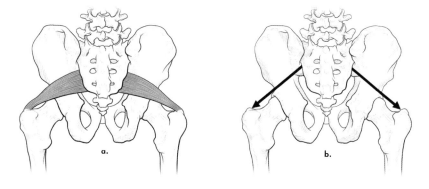

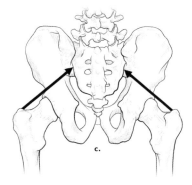

Fig. 11-12. a. The piriformis muscle; b. The piriformis acts on the trochanter; c. The piriformis stabilizes the sacrum.

sacrum is destabilized and also shifts to one side. To counteract this, the piriformis contracts and maintains the stability of the sacrum, thus acting as a stabilizer of the sacroiliac joint and lower spine (Fig. 11-12c). This is particularly true when we stand with our weight shifted over one hip, as many of do habitually, and when we walk or run, in which case the piriformis must alternately release and contract to maintain the stability not only of the hip but of the sacrum as well.

In stabilizing the sacrum and the sacroiliac joints, the piriformis does not simply act on the hip but, taking the hip as a fixed point, it acts in the other direction: on the sacrum. Thus the piriformis muscle performs the basic postural functions of stabilizing both the hip and the sacrum, making it different from all the other deep muscles of the hip, which act in only one direction. The piriformis muscles are unique in one other respect: because they join into a common structure that is constantly shifting, they act on the sacrum from both sides in order to maintain its stability, acting as a unit to stabilize posture.

The problem comes when we habitually shorten the hips and legs, causing the piriformis muscle to overwork in its attempt to stabilize the sacrum and to maintain balance. Over time, the muscle becomes shortened and fixed, and its stabilizing function becomes a liability. To serve its postural function, this muscle—like all the fan muscles of the hip—must release, which can be felt as a kind of widening and toning across the back of the pelvis. The piriformis can then perform its stabilizing and postural function, supporting both the sacroiliac joint and the spine, without compromising the freedom of the hip joint and the ability of the legs to lengthen out of the back. The key to this balanced function, as with all our muscles, is that this muscle must lengthen between its bony contacts and contract, as needed, in the context of this lengthened state.

PIRIFORMIS SYNDROME

Although low back pain is often caused by damage or compression in the vertebrae of the lower back, this is not the only cause of referred pain in the hip and leg. The sciatic nerve, which is the largest nerve tract in the body and which serves much of the leg, passes just underneath the piriformis muscle (Fig. 11-13). If the piriformis muscle has been injured or is chronically contracted, it can pinch or compress this nerve, causing pain in the buttocks region and down the leg. This condition, often referred to as *piriformis syndrome,* often goes undiagnosed because nerve-related symptoms are often attributed to the spine and not the hip region. If the buttocks muscles release and the pain is reduced, however, this indicates that the problem is not related to the lower back but to muscle spasms in the hip region. In any case, not all buttocks and leg pain—which is referred to generically as *sciatica*—originates in the lower back; it sometimes comes from the hips.

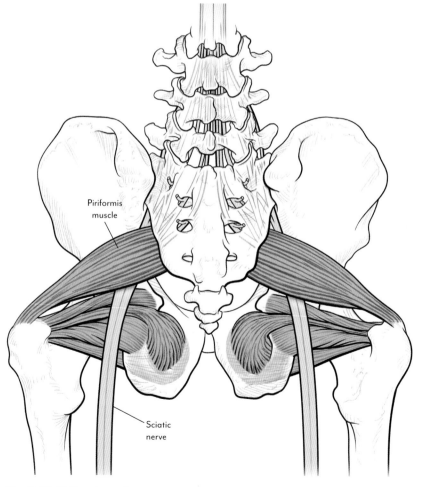

Fig. 11-13. Piriformis syndrome.

THE ROCKERS OF THE PELVIS: OUR "FEET" WHEN SITTING

When the first primates began to walk fully upright, the spine began to function as a vertical, weight-bearing column that transferred the weight of the trunk onto the legs and feet. This made possible an entirely new form of locomotion that was not only efficient in itself but also freed the arms to function in a new way.

But standing and walking are not the only form of upright posture. As long as we are not slumping, we are also fully upright when we sit in a chair, with our weight supported not on the feet but on the ischial tuberosities of the pelvis, or sit bones. As a form of upright posture, sitting has, in many ways, replaced standing. It is in the sitting position that we perform many of our most complex activities, including using our

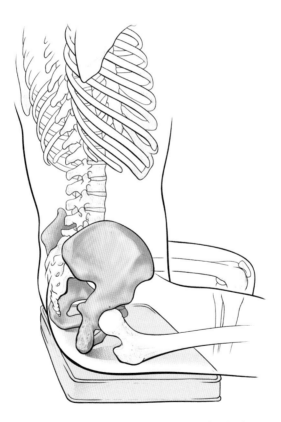

Fig. 11-14. The rockers of the pelvis are our "feet" when sitting.

arms in fine motor tasks such as playing a musical instrument, typing, or using tools to make or fix things. Because sitting in the modern human has become such a universal activity, the chair and table—not to mention couches and other kinds of upholstered furniture—have become dominant features of the human landscape.

But the sit bones are more than mere contacts for sitting on chairs. Although we think of the sit bones as somehow part of the buttocks and pelvis, they function as extensions of the spine. To move on land, four-footed animals need to support the trunk with the limbs. The forelimbs, even in a cat or a dog, need to remain highly mobile in order to absorb shock, which is why they are not directly linked to the spine. Not so with the rear limbs, which need to convey direct thrust to the trunk. To do this, the "scapula" of the pelvis became directly connected to the spine to form the sacroiliac joint, linking the pelvis directly with the spine as a unified bony structure.

If this link is critical in four-footed animals, it becomes in some ways even more so in humans. When we stand, the weight of the trunk is transferred from the spine directly to the pelvis, and from there onto the heads of the femurs so that, for all intents and purposes, the pelvis becomes an extension of the spine and even, in some ways, part of the trunk. This linkage relates directly to sitting. When we sit, we quite literally rest the trunk on our sit bones, which are not mere projections at the bottom of the pelvis but extensions of the spine. When we sit, we support weight not on our feet but on the rockers of the pelvis, which are our "feet" when we are sitting (Fig. 11-14).

THE PELVIS AND THE TRUNK

When asked where the hip joints are located, many people point to the pelvic bones—the bones at the top of your pelvis that form your waist. Technically speaking, these are not your hips but the crest of the pelvic or iliac bones; the word *hip*, as any doctor will tell you, refers not to the iliac crest but to the hip joint, which is much lower than the upper rim of the pelvic bones. The confusion between the two is echoed in the linguistic similarity between the names we use for these two very different parts of the pelvis (*hips* versus *hip joints*) (Fig. 11-15).

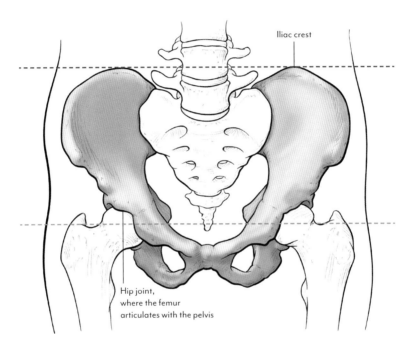

Iliac crest

Hip joint,
where the femur
articulates with the pelvis

Fig. 11-15. The hips (red) versus the hip joints (green).

In practical terms, this misconception translates into the tendency, when bending or sitting, to create a joint at the waist and to shorten and collapse. To maintain the proper length of the trunk when bending, we must appreciate that the pelvis functions as part of the back, and that bending or flexing the trunk should take place not at the waist but at the hip joints (Fig. 11-16a and b). Maintaining the full length of the trunk requires that we know where the hip joints are located and how to use them properly.

There is a corollary to this principle of understanding where the hip joints are located. If the pelvis is connected to the spine, and if bending takes place at the hip joints, then the back of the pelvis, or sacral region of the spine, functions not as part of the legs but as part of the length of the back (Fig. 11-16c). In Chapter 3, we saw that the sacrospinalis muscles extend from the sacrum right up to the base of the skull, and the sacrum provides a stable point of origin for the extensor muscles running up the length of the back. Because the sacrum is attached to the pelvis, the pelvis is more than a stable support structure for the legs and its muscular attachments; it is, in fact, part of the back and must be thought of in this way if we are to maintain a fully lengthened trunk.

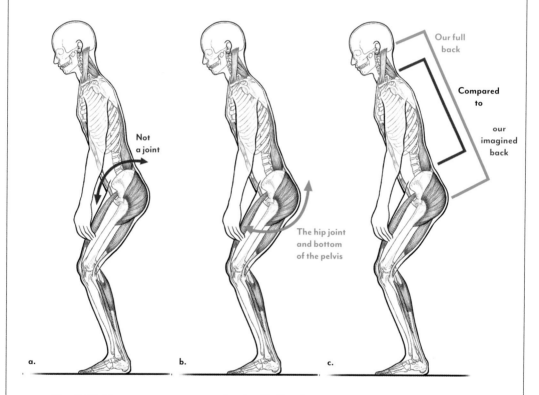

Fig. 11-16. The way we perceive our back will affect how we use it.

EXERCISE:
"Monkey" Position with Sit Bones on a Table Surface

To experience how we flex at the hip joints, let's take the previous monkey position exercise from Chapter 5 and add another step.

1. Stand with your back to a table that goes no higher than the tops of your legs. Place your feet about shoulder width apart so that you can bend at your knees without jamming your hips. Be aware of your feet on the ground and, without lifting your chest up or puffing it out, come up to your full stature.

2. Think of releasing the knee joints; allow your knees to bend and lower your entire trunk in space until your buttocks contact the table. Let your pelvis come back slightly as you adjust your weight evenly on the front and back of your feet. In this vertical monkey position, with your trunk lowered in space, your sit bones (that's the bottom of your pelvis) should be very nearly, if not actually, in contact with the table.

3. Notice how, in this position, your sit bones clearly function as the bottom end of your back and trunk. In this position, you should be able to very clearly feel the entire length of your back and have a much fuller kinesthetic sense of how long the back really is.

4. Incline forward carefully, in such a way that you do not lose the length of your back by bending at your "false hips." In this position, you can see clearly that, when you bend at the hips, you are bending not at the waist but are hinging at the hip joints, with the pelvis clearly functioning as part of the trunk and back and not the legs.

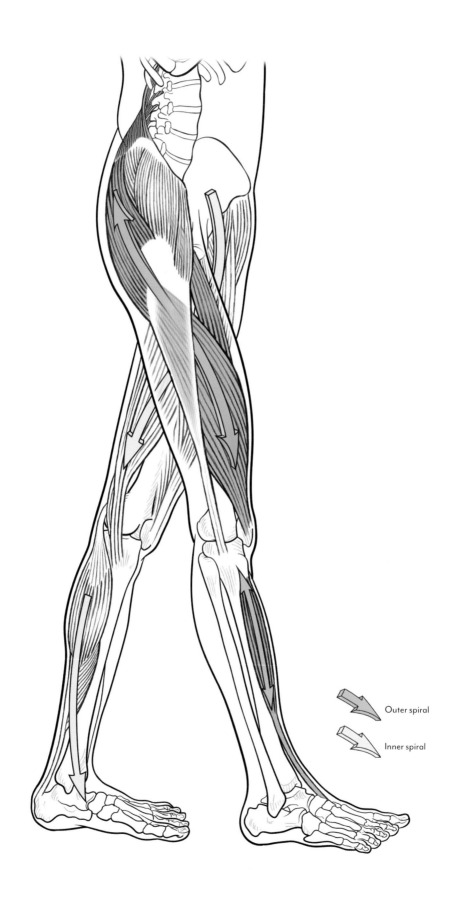

Outer spiral

Inner spiral

12

The Leg Spirals

The lower limb is a system of levers for support and propulsion on two feet. In contrast to the upper limb, which is modified for grasping and manipulating objects, the legs are designed for locomotion on two legs. For this purpose, the legs are much sturdier than the arms, carrying the weight directly above the ground by means of bones that are stacked on top of one another and ultimately rest on the arched foot.

The Key to the Leg Spirals

In order to function efficiently, the legs must maintain the support of the trunk with minimum effort and maximum mobility. This requires the muscles of the hips, thighs, and lower legs to maintain length while ensuring the freedom and mobility of the joints.

Contraindications

Many of us stand by habitually bracing the legs. The long muscles of the legs are overworked, the knees are often hyperextended, and the lower leg and ankle are fixed, even while walking. Instead of lengthening against gravity, the legs become shortened and overworked.

WHEN WE BRACE AND STIFFEN THE LEGS

1. The gluteal muscles are overworked and the hip joints are fixed.
2. The quadriceps are tight and overworked.
3. The adductors and hamstrings are shortened.
4. The gastrocnemius and soleus muscles are shortened and overworked.
5. The lower leg is tight and the ankle and foot joints are fixed.

WHEN THE LEGS LENGTHEN

1. The hip and buttocks muscles release to free the hip joints.
2. The quadriceps muscles soften and release.
3. The adductors and hamstrings lengthen.
4. The gastrocnemius and soleus muscles release and the heels go onto the ground.
5. The muscles of the lower leg release so that the ankle releases.
6. The foot fully contacts the ground.

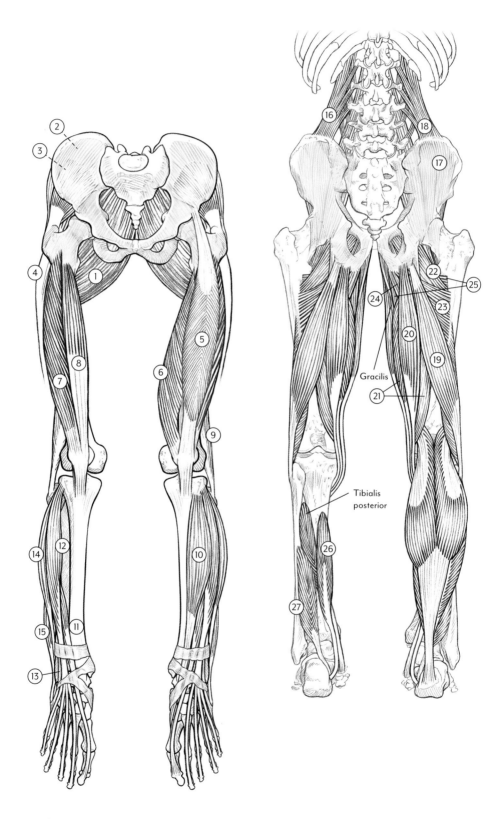

Gracilis

Tibialis posterior

ORIGIN	MUSCLE	INSERTION
OUTER (EXTENSOR) SPIRAL		
Ilium/sacrum	**1. Gluteus maximus**	Iliotibial (IT) band/shaft of femur
Inner border of ilium	**2. Gluteus medius**	Greater trochanter
Surface of ilium	**3. Gluteus minimus**	Greater trochanter
Anterior iliac crest	**4. Tensor fascia latae**	IT band
Ilium above acetabulum	**5. Rectus femoris**	Tibial tuberosity
Upper shaft of femur	**6. Vastus medialis**	Tibial tuberosity
Upper shaft of femur	**7. Vastus lateralis**	Tibial tuberosity
Upper shaft of femur	**8. Vastus intermedius**	Tibial tuberosity
Lower shaft of femur	**9. Biceps short head**	Head of fibula
Lateral tuberosity/upper shaft of tibia	**10. Tibialis anterior**	Cuneiform/first metatarsal
Fibula	**11. Extensor hallucis longus**	Big toe
Shaft of tibia/lateral condyle of tibia	**12. Extensor digitorum longus**	Four toes
Fibula	**13. Peroneus tertius**	Fifth metatarsal
Upper fibula	**14. Peroneus longus**	First metatarsal
Lower fibula	**15. Peroneus brevis**	Metatarsal of the little toe
INNER (FLEXOR) SPIRAL		
Iliac crest	**16. Quadratus lumborum**	L1–4, T12
Anterior surface of ilium	**17. Iliacus**	Lesser trochanter
T12, L1–L4	**18. Psoas major**	Lesser trochanter
Ischium	**19. Biceps femoris**	Head of fibula
Ischium	**20. Semitendinosus**	Below medial epicondyle of tibia
Ischium	**21. Semimembranosus**	Medial epicondyle of tibia
Pubic bone	**22. Pectineus**	Upper shaft of femur
Pubic bone	**23. Adductor brevis**	Upper shaft of femur
Pubic bone	**24. Adductor longus**	Middle shaft of femur
Pubic bone/ischium	**25. Adductor magnus**	Medial femoral condyle
Middle of tibia	**26. Flexor digitorum longus**	Lateral four toes
Lower fibula	**27. Flexor hallucis longus**	Big toe

The Leg

Like the upper limb, the leg is composed of one long bone (the femur) and two lower leg bones (the tibia and fibula), a cluster of tarsals, or ankle bones, and the bones of the foot (the metatarsals and phalanges)—three for each little toe and two for the big toe (Fig. 12-1). As with the upper limb, the long bone of the leg forms a ball-and-socket joint with its supporting girdle. The two segments of the leg hinge at the knee, and the foot can be moved and adjusted in relation to the lower leg, not for the purpose of grasping objects (the big toe has lost the ability to oppose the other toes), but in order to adjust the foot to the ground for balance and movement.

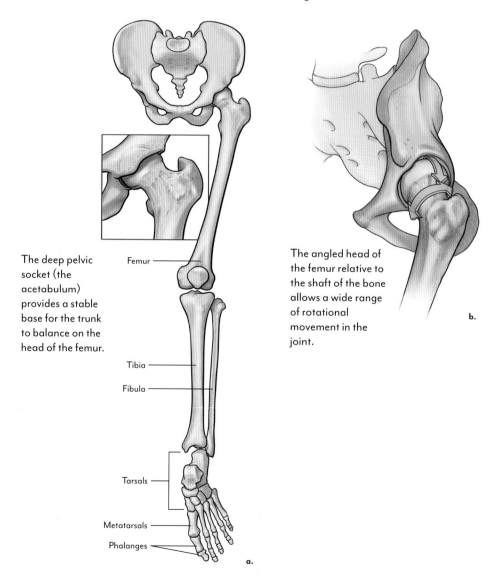

The deep pelvic socket (the acetabulum) provides a stable base for the trunk to balance on the head of the femur.

Femur

Tibia

Fibula

Tarsals

Metatarsals

Phalanges

The angled head of the femur relative to the shaft of the bone allows a wide range of rotational movement in the joint.

a.

b.

Fig. 12-1. Bones of the left lower limb showing: a. The distinctive arrangement of bones also found in the upper limb; b. The detail of the hip joint.

One of the key functions of the lower limbs is to flex and extend the legs, both in walking and in lowering and raising the trunk. To support ourselves against gravity, we must be able to maintain extension at the ankles, knees, and hips, but in such a way that the joints are not braced or stiffened (Fig. 12-2). This makes it possible for the knees to bend easily when we need them to, and for the leg muscles, which are capable of very powerful movements, to maintain background postural support with a minimum of effort and strain.

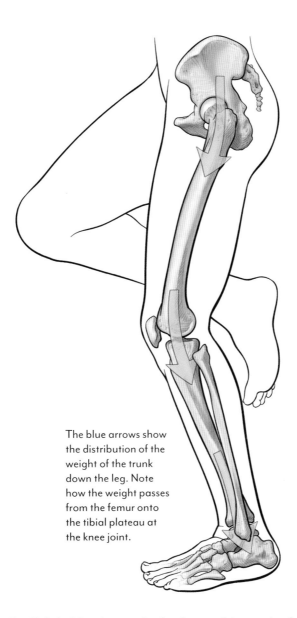

The blue arrows show the distribution of the weight of the trunk down the leg. Note how the weight passes from the femur onto the tibial plateau at the knee joint.

Fig. 12-2. Left leg showing the distribution of the weight of the trunk down the leg. The limb transmits the gravitational force of the body onto the ground and moves freely at the joints.

The legs are also capable of spiraling movements, as can be easily seen if you look, say, to your right while standing and then begin to turn your body in that direction. As your trunk rotates in the direction of your gaze, the heel of your left leg begins to come up off the ground as that knee bends, allowing the leg to swing under you so that you begin to walk to your right. All of this is based on torsional movement not only of the trunk, but also of the legs, from the feet up to the pelvis, as the legs are organized in spirals just as the trunk is. To fully understand how to use the legs without bracing and stiffening them, we need to understand not only how we flex and extend the limbs, but also how these spiral muscles work.

A Tale of Two Spirals

A simple way to understand the spirals of the legs is to observe a child crawling on the floor. At this stage of development, a child is rather like a reptile with limbs splayed out to the sides. To move along the ground, the child first advances a knee and then, to propel itself, pushes off with the knee and foot by extending the leg. These are the two spiral movements of the leg. To advance the knee, the leg rotates outwardly (Fig. 12-3a); when the knee is extended the leg rotates inwardly (Fig. 12-3b). The first movement corresponds to the flexor (or inner) spiral that crosses the front of the hip and continues down the inner thigh and the back of the lower leg; the second movement corresponds to the extensor (or outer) spiral that begins with the gluteal muscles at the back of the hip and wraps around to the quadriceps muscles on the front of the thigh and down the front of the lower leg (Fig. 12-4).

a.

b.

Fig. 12-3. a. When an infant crawls, the advancing limb rotates outward; b. The extended limb rotates inward.

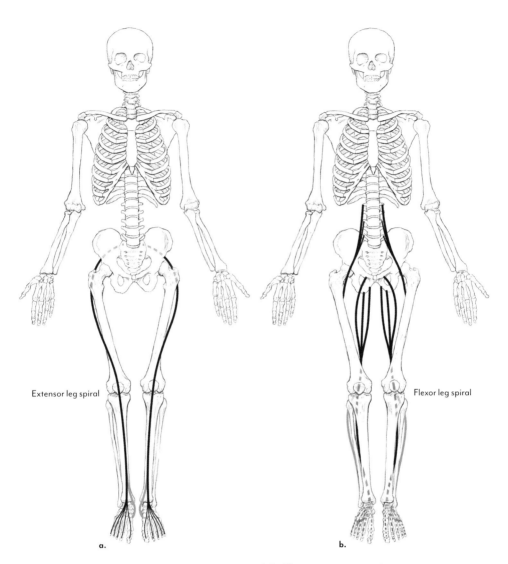

Fig. 12-4. The leg spirals: a. Extensor or outer spiral; b. Flexor or inner spiral.

The crawling child in Fig. 12-3 is representative of a developmentally primitive example of the spirals, but these actions take place even when we are walking fully upright. As we bend a knee to advance the leg, the leg rotates outward; as we extend the leg, it rotates inward. In the crawling child, these movements take place quite effortlessly, but in the average adult, each of these spirals is prone to shortening, which contributes to the bracing of the legs. If you look at most people when they stand, you can see that this second action of extending the limb is exaggerated so that the thigh is braced and shortened. We can see this tendency even more clearly in people who, as they get older, get so stiff in the legs that they have trouble bending at the knees to sit and stand (Fig. 12-5). The cause of this stiffness becomes clear when we look at the line of the extensor muscles—particularly vastus intermedius, vastus lateralis, and the iliotibial tract (or IT) band. The IT band is a very strong, fibrous band of connective tissue that originates on the side of the hip at the greater trochanter and runs down the side of the leg to the outer knee; it is mostly fibrous, with a muscular part, the tensor fascia latae, at the top. The two vastus muscles, which are very powerful and exert strong pulls, originate at the greater trochanter of the hip and, angling inward, attach at the extensor tendon that joins into the kneecap. When these muscles are tight, they extend the leg and the knee and pull the knee outward (Fig. 12-6).

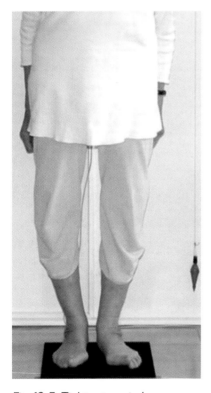

Fig. 12-5. Tight outer spiral.

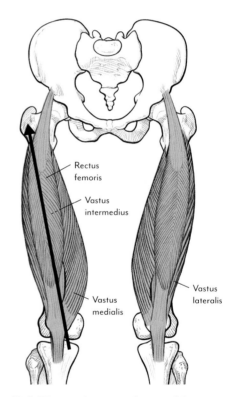

Fig. 12-6. The quadriceps at the top of the outer spiral showing the line of pull.

The second tendency is to shorten the inner thighs. Most of the muscles of the inner thigh, which originate at the pubic bone, attach at the inner knee or along the femur in the general direction of the inner knee (Fig. 12-7). Two of the hamstrings also attach to the inner condyle of the tibia. The line of pull of these muscles is very clear: they pull the knee in, and they pull the crotch into the knee, causing a shortening of the inner thigh that results in a loss of postural support in the legs (Fig. 12-8).

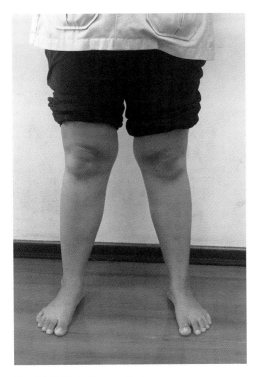

Fig. 12-7. Tight inner spiral.

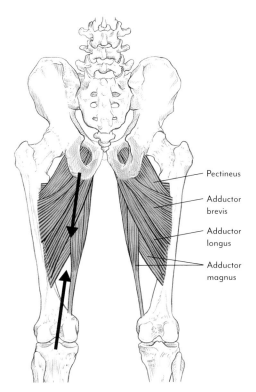

Pectineus

Adductor brevis

Adductor longus

Adductor magnus

Fig. 12-8. Adductors of the inner spiral and the line of pull (dorsal view).

THE THREE MUSCLE COMPARTMENTS OF THE THIGH

There are three groups of muscles on the thigh: the flexors, or hamstrings, on the back of the legs; the adductors on the inside of the thigh; and the extensors, or quadriceps group, on the front of the thigh; each group is contained within a separate compartment. Viewed in cross-section, we can clearly see the three compartments: the hamstrings in blue, the adductors in purple, and the quadriceps—the largest by far—in green (Fig. 12.9). The hamstrings and adductors are flexors and are part of the inner spiral; the quadriceps make up the outer spiral. All these muscles powerfully move the leg at the knee and hip, but they are also postural in function and, to work properly, must maintain the weight of the trunk without shortening. When we send or direct the knees out of the trunk, it is these three muscle groups that must respond to our directive thought, releasing and toning until they begin to let go out of the back and pelvis. This restores the lengthened support of the legs, which can be felt as a kind of springiness and pliant tone in the legs.

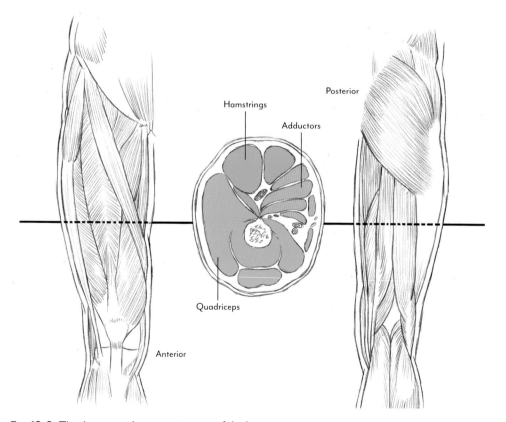

Fig. 12-9. The three muscle compartments of the leg.

Tracing the Leg Spirals

Let's look now at the two spirals of the legs in greater detail, beginning with the front or outer spiral that originates at the gluteal muscles and continues down the thigh, lower leg, and into the foot.

The Outer Spiral

The outer spiral is most obvious in the line of the extensor muscles of the thigh (Fig. 12-10a).

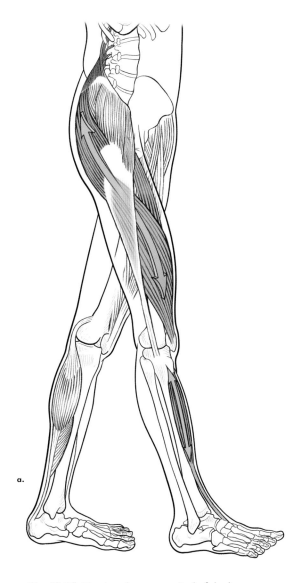

Fig. 12-10. Tracing the outer spiral of the leg:
a. The outer spiral (in blue) runs the length of the leg.

The spiral itself begins with the gluteal muscles (Fig. 12-10b). These muscles originate at the iliac crest and sacrum and run obliquely downward to the greater trochanter of the hip. This line continues with the extensors of the thigh, two of which clearly angle inward to the knee and attach to the tendon of the kneecap. From here, we can continue to trace this spiral with the anterior muscles of the lower leg, which taper into tendons that cross the top of the ankle to extend the toes (Fig. 12-10c).

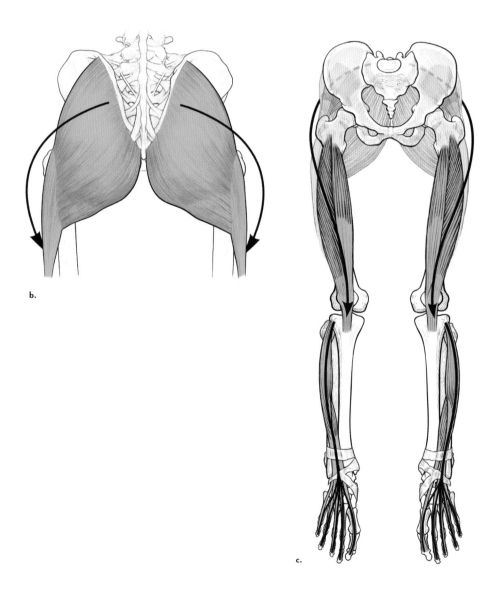

b.

c.

Fig. 12-10. Tracing the outer spiral of the leg (cont.): b. The spiral begins with the gluteal muscles, which wrap around from the sacrum in back to the outer thigh; c. The spiral continues with the quadriceps muscles, which follow the inward angle of the femur from the upper thigh to the knee and continue to the dorsal surface of the foot.

The Second "Subsidiary" Outer Spiral

A second spiral line follows the IT band to tibialis anterior, which crosses the front of the lower leg to the inner ankle, wraps underneath the foot, and attaches to the little toe (Fig. 12-11). At the little toe joint, it meets the third outer spiral.

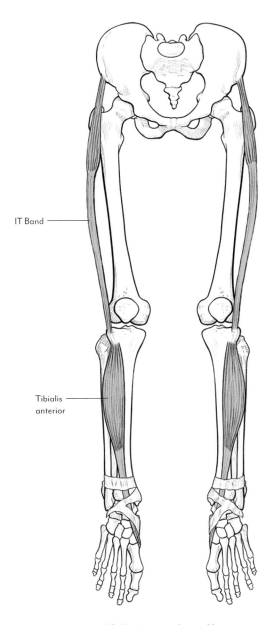

IT Band

Tibialis anterior

Fig. 12-11. A second spiral line extends from the hip down the IT band to tibialis anterior through to the inner ankle.

The Third "Ambiguous" Outer Spiral

The third outer spiral follows the IT band to peroneus longus, which goes around the outside of the foot and then underneath to attach to the big toe (Fig. 12-12a). Acting together, peroneus longus and tibialis anterior form a stirrup underneath the foot (Fig. 12-12b). We will see later in this chapter that this spiral has the ambiguous characteristic of sharing a muscle—the peroneus longus—with an inner spiral.

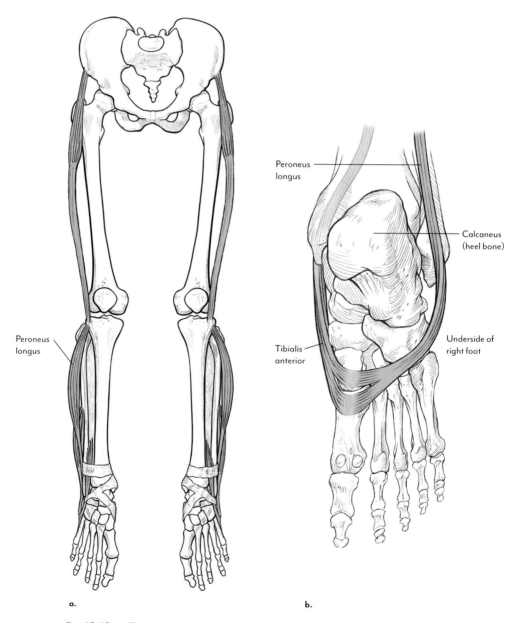

Peroneus longus

Peroneus longus

Peroneus longus

Calcaneus (heel bone)

Tibialis anterior

Underside of right foot

a.   b.

Fig. 12-12. a. The third spiral line travels down the IT band and crosses into the peroneus longus to travel to the outside of the ankle; b. Peroneus longus and tibialis anterior form a stirrup underneath the foot joining the second and third outer spirals.

The Inner Spiral

The flexor, or inner spiral (Fig. 12-13a), begins at the back of the pelvis with the quadratus lumborum and the iliopsoas muscles, the key flexors of the hip (Fig. 12-13b).

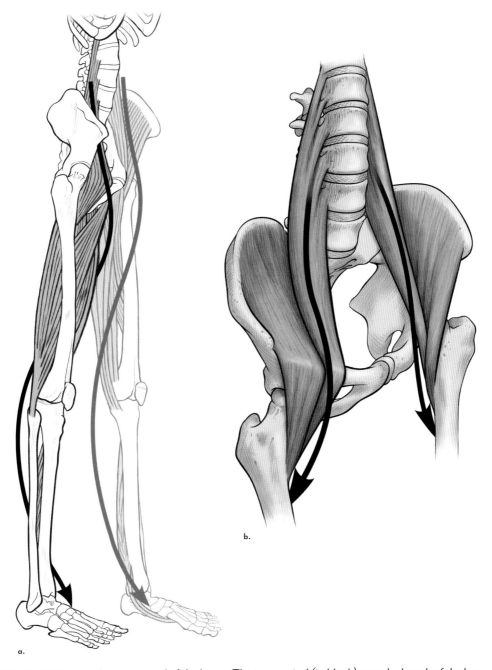

Fig. 12-13. Tracing the inner spiral of the leg: a. The inner spiral (in black) runs the length of the leg; b. The spiral begins with the psoas muscles, the key flexors on the front of the pelvis, which originate on the anterior spine and insert into the inner femur.

From here, the flexor line continues with the adductors and hamstrings, verging mainly toward the inner knee. This line continues with the deep flexors on the back of the lower leg (flexor digitorum longus, flexor hallucis longus, and tibias posterior) that flex the toes (Fig. 12-13c).

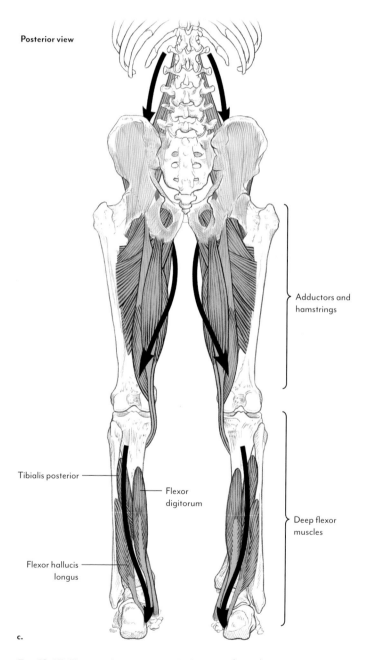

Posterior view

Adductors and hamstrings

Tibialis posterior

Flexor digitorum

Deep flexor muscles

Flexor hallucis longus

c.

Fig. 12-13. Tracing the inner spiral of the leg (cont.): c. The spiral continues with the adductors of the inner thigh and the hamstrings, which attach to the back of the knee. The spiral continues with the deep flexors on the back of the lower leg that attach into the foot.

The Subsidiary Inner Spiral

We can trace a subsidiary spiral if we follow the biceps femoris muscle from its origin at the ischium to the outer knee. From here, we can continue with peroneus longus to the outer foot and follow this under the foot to the big toe (Fig. 12-14). This spiral has the same ambiguity as the third outer spiral, as it also employs the peroneus longus muscle.

As we saw, the tendency to shorten these muscles causes us to sink into and stiffen the legs so that we are not able to flex fully at the knees and, instead, brace the legs when we stand, seizing up in the hips and around the gluteal muscles.

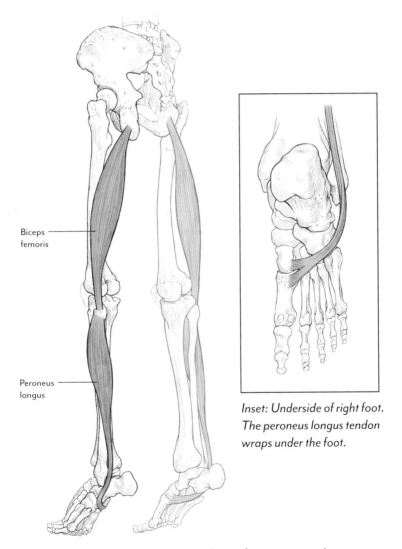

Biceps
femoris

Peroneus
longus

Inset: Underside of right foot. The peroneus longus tendon wraps under the foot.

Fig. 12-14. Alternate inner spiral: Biceps femoris to outer knee; peroneus longus to the outer foot—note that in Fig. 12-12a, the peroneus longus also forms one of the outer spirals.

EMBRYOLOGICAL ORIGINS OF THE LOWER LIMB SPIRAL

In Chapter 10, we saw how the limbs in the developing embryo rotate to produce spiral patterns, the upper limb rotating outwardly and the lower limb rotating inwardly. Both of these rotations are directly related to our human upright posture. When primates began to stand on two feet, the upper limb assumed a prehensile function. In this new upright posture, the trunk and chest flattened and the shoulders extended more to the sides. This is an important reason why the upper limb rotates outward: to produce the distinctive, widened shoulder girdle that organizes the trunk more closely around the spine and supports the upper limb in a new way. At the same time that the upper arm rotated outward, the hand pronated, producing the spiral of the upper limb.

To support upright posture, the legs rotate inward. To stand on two feet, the legs must assume the entire weight of the trunk; to do this, they must not only be brought under the body but be arranged vertically, with the upper leg stacked on the lower. For this to happen, the leg rotates medially (Fig. 12-15b and d), bringing the extensor muscles on the back part of the leg around to the front so that these extensors can maintain extension of the leg at the knee and support fully upright posture.

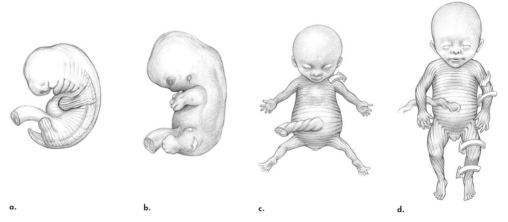

a.   b.   c.   d.

Fig. 12-15. Embryological origin of the spirals of the lower limbs.

This rotation explains why the extensors of the leg, in contrast to the extensors of the neck and trunk, are on the front of the body and not in back. The purpose of the extensors of the neck and back is to extend the head and trunk against gravity or, to put it differently, to keep them from buckling. This requires muscles at the nape of the neck, along the whole of the trunk, and at the hip—in other words, in back.

Maintaining extension of the leg at the knee, however, requires muscles on the front of the thigh—hence the exception to the rule. As with the spirals of the upper limb, this spiral pattern can be clearly seen in the dermatomes, which spiral from the back around to the front (Fig. 12-16).

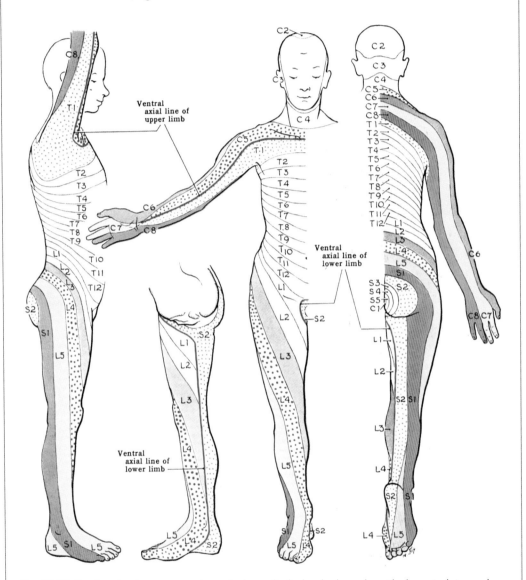

Fig. 12-16. Tracing the dermatomes down the lower limb clearly shows how the leg spirals inward.

The Tendons Acting on the Foot

On the forearm, two groups of muscles taper into tendons that cross the wrist; four groups of tendons cross the ankle. First, the gastrocnemius and soleus muscles (and the lesser-known plantaris muscle) converge into the Achilles tendon, which inserts directly into the calcaneus, or heel bone (Fig. 12-17a). The heel bone juts out from the ankle to form a lever, so these muscles, acting on this lever, powerfully plantar flex the foot at the ankle (Fig. 12-17b).

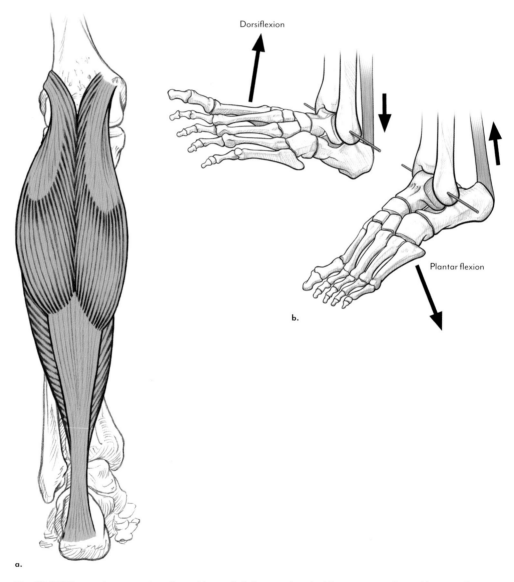

Fig. 12-17. The tendons crossing the ankle: a. Achilles tendon; b. Movement at the ankle joint showing the action of gastrocnemius.

Second, the peroneal muscles cross the outside of the ankle and attach to the side and underside of the foot, everting the foot (Fig. 12-18a). Third, the extensors of the foot cross the front of the ankle to attach to the dorsal surface of the toes, extending the foot and toes (Fig. 12-18b). Fourth, the deep flexors of the lower leg taper into tendons that cross behind the medial malleolus and attach underneath the foot, flexing the toes and inverting the foot (Fig. 12-18c).

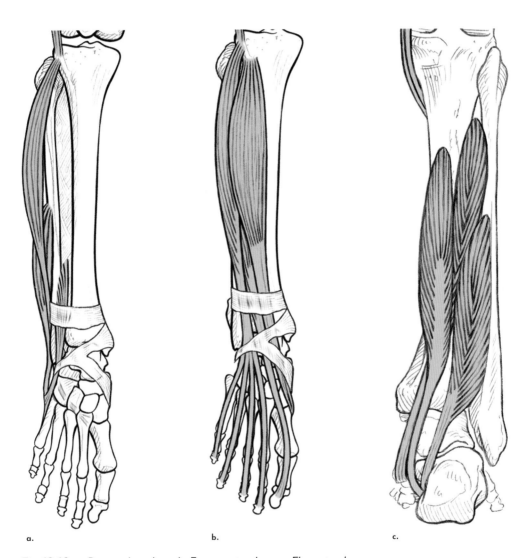

a. b. c.

Fig. 12-18. a. Peroneal tendons; b. Extensor tendons; c. Flexor tendons.

WHERE IS THE WRIST OF YOUR FOOT?

On your forearm, the flexors taper into tendons that cross the carpal tunnel and attach to the palmar side of the wrist and hand; the extensor tendons cross the back of the wrist and attach to the carpal bones and fingers on the dorsal side of the hand. Like the forearm, the extensors of the foot cross the front of the ankle to attach to the dorsal surface of the toes; this part of the ankle is equivalent to the outside of your wrist (Fig. 12-19a). But where are the flexors, and where do they cross the ankle? Where you might expect to find the flexors—that is, in back—the gastrocnemius and soleus muscles (and the lesser-known plantaris muscle) converge into the Achilles tendon, which inserts directly into the calcaneus, or heel bone. These move the foot at the ankle but do not attach to or flex the toes.

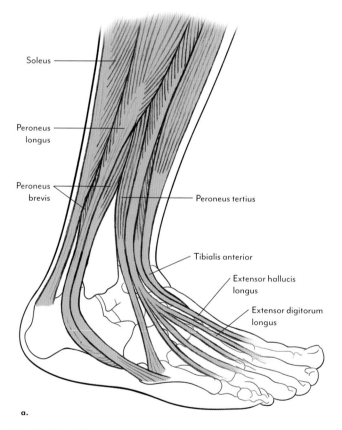

Soleus

Peroneus
longus

Peroneus
brevis

Peroneus tertius

Tibialis anterior

Extensor hallucis
longus

Extensor digitorum
longus

a.

Fig. 12-19. a. The extensor tendons pass over the front of the leg and foot and the peroneal tendons, which evert the foot, run down the outside of the leg.

So where are the deeper flexors that act on the toes and invert the foot, and where do they cross the ankle? These tendons cross behind the medial malleolus and sneak under the arch of the foot to attach onto the plantar surface of the foot; this area just behind the medial malleolus on the ankle is the "wrist" of the foot (Fig. 12-19b).

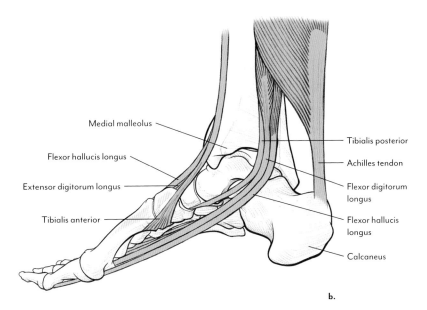

Fig. 12-19. (cont.): b. The deep flexors on the inside of the leg cross behind the medial malleolus and invert the foot.

The Lower Leg and the Transverse Joint of the Foot

In order to position the foot in relation to the ground, we are able not only to hinge the foot at the ankle but to evert and invert the forefoot (Fig. 12-20). This action is crucial because, if you can only hinge the foot at the ankle, the foot is not able to maintain its contact with the ground when confronted with uneven terrain or changes in the position of the leg (as when pushing off one foot). The main muscles that produce eversion are the peroneal muscles on the outside of the leg; inversion is produced by the flexors of the lower leg (Fig. 12-19a and b). Because we tend to overuse these flexors, most of us habitually flex the toes and invert the foot, fixing the joint and causing the foot as a whole to lose flexibility over time. When the flexors release, the joint becomes freer, the toes can lengthen out of the plantar surface of the foot, and the foot as a whole opens onto the ground.

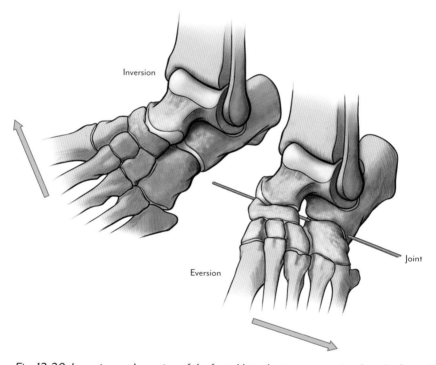

Inversion

Eversion

Joint

Fig. 12-20. Inversion and eversion of the foot. Note that movement involves the frontal five tarsal bones moving around the first two that are static, forming a joint.

OUR ANKLE JOINT AND HOW WE ARE POISED ON TWO FEET

The most obvious function of the ankle is to move and position the foot in relation to the lower leg by dorsiflexing and plantar flexing the foot. When we raise our weight on our toes or push off the ground when walking, for instance, we are plantar flexing the foot. We can also see that, when the foot is fixed on the ground, the muscles on the lower leg that act on the foot now stabilize the leg in relation to the foot, which is essential to our postural support and balance on two feet.

But these actions represent only a part of what the ankle is for. In our upright vertical posture, the entire body is poised on two feet as the basis for moving in space. When we walk, we bend a knee to advance a leg but, to take this step, we must first incline at the ankle in order to initiate the forward movement of the body as a whole (Fig. 12-21). In this case, the leg, and the entire body with it, must be free to incline at the ankle, and this requires that the lower leg move freely at the ankle. Although we must stabilize the lower leg in relation to the foot as the basis for maintaining upright posture, we must also be unstable in order for movement to take place, and this instability is actually part of how we are designed to move on two feet. If we shorten in the legs and brace the knees, this interferes with our ability to move freely at the ankle and thus compromises our upright poise. Freedom at the ankle joint is thus an essential component of our upright design.

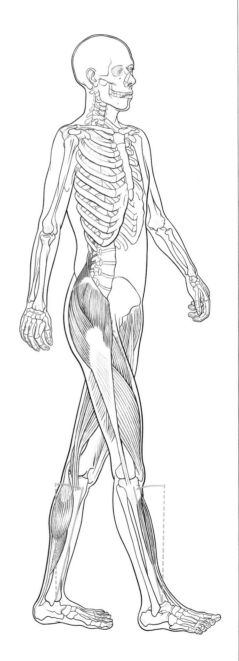

Fig. 12-21. With the foot on the floor and the lower leg pivoting at the ankle joint, the body as a whole is poised on two feet.

THE ANKLE AND THE LEG

When we brace and stiffen in our legs, two regions that are clearly implicated are the adductors and extensors of the thighs, which in most of us become chronically tense and overworked. But the lower leg also plays a key role in how the legs work, often becoming chronically tight and rigid. Because the muscles of the lower leg act on the foot and ankle, tightening these muscles means that we are holding in the ankle and pulling up the foot—something we can see quite clearly in the gastrocnemius muscle, which becomes overly contracted, raises the heel off the ground, and stiffens the ankle. This is why it is so important to stop bracing the back of the knees and to allow the back of the legs to lengthen, which in turn releases the heels onto the ground (Fig. 12-22).

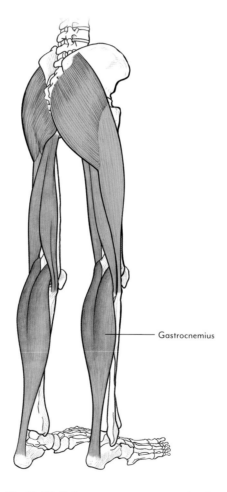

Gastrocnemius

Fig. 12-22. If the gastrocnemius muscles shorten, the heel is pulled off the ground (see Fig. 12-17).

As important as the back of the leg is, however, this is not the only area of the leg that needs to let go, nor is it the key to the ankle joint, which we must look at further to make sense of this region. It is interesting to note that the Achilles tendon and calcaneus are not even directly involved in the ankle joint; as a kind of lever arm, the calcaneus extends backward from the joint, so that the gastrocnemius, pulling on the calcaneus, acts on the joint but is not intimately a part of it. To understand the ankle joint, we must look to a key group of muscles that act on the ankle, and the complex of tendons that cross the front of the ankle joint: the extensors of the leg. We've seen that there are three extensors of the lower leg: extensor digitorum longus, extensor hallucis longus, and tibialis anterior. All three of these muscles originate at the upper tibia and fibula and, descending to the ankle, cross over the front of the ankle and onto the dorsal surface of the foot, where the larger tendon breaks into sections that attach to all the toes, including the big toe (Fig. 12-23). When we habitually shorten in the lower leg, we tighten and brace this part of the ankle—something we can see when someone loses their balance and raises their toes to keep from falling backward. In such cases, we are quite literally pulling the forefoot off the ground in an effort to maintain our balance; in the attempt to pull the tibia forward, we also pull the foot up. All of this is not only unnecessary but positively harmful to our balance and poise on two feet. For the legs to work properly, the ankles need to let go so that the foot can fully contact the floor and, in turn, we can lengthen up from the foot (Fig. 12-24).

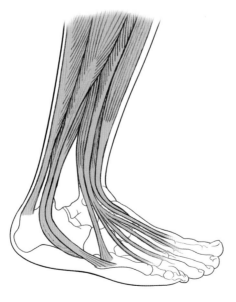

Fig. 12-23. Bracing the extrinsic muscles of the foot, situated in the lower leg, prevents the foot releasing onto the ground.

One way to make sense of this region is to think of the front of the ankle as the "nape of the neck," the foot as the "head," and the lower leg as the "trunk"—a kind of upside-down head, neck, and trunk. If we tighten the neck/ankle and pull back the head/foot, we are shortening in the trunk/lower leg. To lengthen the system, we need to release the front of the ankle to let the foot go onto the ground; this allows the lower leg as a whole to lengthen. Freeing the front of the ankle is thus key to releasing and lengthening the lower leg muscles.

It may seem somewhat counterintuitive to suggest that releasing the ankles will help the heels go onto the ground when, in fact, the opposite seems to be the case: if we tighten these muscles to dorsiflex the foot, doesn't this put the heel onto the ground? The answer is that, if the muscles crossing the front of the ankle are tightening, they are pulling the foot up and, in turn, preventing the lower leg from lengthening. When the front of the ankle lets go (indicated by the arrow in Fig. 12-24), this allows the entire foot, heel included, to go onto the floor. Allowing the front of the ankle to release is the key to letting go through the leg and letting the feet open onto the ground, which in turn means that we get support of the body from the ground with the knees going away and the legs lengthening—one of the keys to the spiral musculature of the leg.

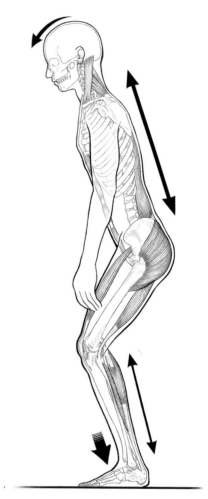

Fig. 12-24. Releasing the front of the ankle allows the foot to fully contact the floor; this in turn allows the leg as a whole to lengthen as part of the activation of the entire postural system.

APPENDIX

Throughout this book I have emphasized that the musculoskeletal system is designed to function as a coordinated whole, and that understanding how this system works, and not specific methods for stretching or strengthening muscles, is the key to natural musculoskeletal function and health. We've also seen that in order to restore the natural, coordinated working of the musculoskeletal system described in this book, it is necessary to holistically "direct" parts of the body in supported positions that enable muscles to regain their natural length and that activate the body's natural upright support system. Over time, one can begin to experience the release of chronically tense myofascial tissue, heightened kinesthetic awareness of the body's natural support, and general improvements in postural and muscular function including breathing, vocal health, muscle tone, and the working of various systems such as the back and limbs. To enable those readers interested in exploring this process, I include an extract from my book *Neurodynamics,* in which I outline the various elements involved in this practice.

"Directing" and the Postural System

Because the body is an architectural system in which muscles relate to body parts organized in the field of gravity, one of the most useful things we can do to bring about the natural working of the muscular system is to place the body in a position that will provide optimal advantage for muscles to let go and for the system to begin to work properly as a whole. The simplest and most basic way to do this is to lie down in the semi-supine position that we have visited throughout this book. This provides almost complete body support while giving the neck, back, shoulders, and leg muscles a chance to lengthen and release. Keep in mind, however, that lying down in this way must not be used as a relaxation exercise but as a way of bringing about tone and release in muscles that are habitually contracted. This requires energy and not collapse, and, although it results in a lowered state of nervous energy, it must also be associated with a heightened state of alertness, muscle tone, and vitality. A corollary of this principle is that you must not lie on a cushy mattress or exercise mat, which encourages collapse, but on a fairly hard surface such as a carpeted floor, which supports the body properly.

Three key areas of the body bear weight in the semi-supine position: the occiput, or back of the skull, the scapulae and upper back, and the back of the pelvis. If we are lying on a fairly hard surface, the contact of these weight-bearing areas with the ground will help to stimulate natural postural reflexes. The fixation of these points by gravity also enables us to let go of the muscles of the back, to untwist in the trunk, ribs, and neck, and to let go in the leg muscles.

But this will not happen immediately; we must wait patiently and give the muscles a chance to release. We've seen that since muscles are meant to work elastically and we tend to shorten them, we have to direct the body parts in such a way that the muscles let go and the parts move away from each other. But what do we mean by releasing muscles? Virtually all methods for training, releasing, or stretching muscles attempt, in one way or another, to manipulate or regulate muscle tension, reflecting a belief that muscles are tight and need releasing or stretching, or that they are relaxed and weak and need strengthening. But as we've seen all along, this belief, while based on some truth, is fundamentally wrong in one respect: We can't know how much we should tighten or relax muscles because their length is determined by the natural design of the musculoskeletal system.

When we direct body parts, then, the first thing we must do is stop worrying about muscle tension and think instead about the relations of body parts. Even when we are overly relaxed, muscles tend to shorten and become chronically tight, and for this reason we want to stop tightening, not by thinking directly about muscles but by thinking about the parts of the body that the muscles connect to. We also do not want to focus on any one area but rather on the key areas that make up an organized whole. In particular, we don't want to tighten the neck so that the head can come out of the back; we don't want to arch or narrow the back so that it can lengthen and widen; and we don't want to tighten the legs so that the knees can go forward and away. And we do want to go around the entire circuit so that we don't get stuck on any one part of the body, but think of each piece only in the context of a total coordinated whole.

Now there is one part of these instructions that directly involves muscles, and that is the first instruction for the neck to be free, which focuses on the muscles of the neck. But this instruction should not be interpreted as a direct attempt to relax these muscles but simply as a wish to stop tightening them so that they can relax indirectly on their own. The direction to "free the neck," in other words, is preventive—that is, it is intended as an instruction to stop doing something that interferes with how the body works. Which brings us back to the idea that the directions are, first and foremost, preventive—that is, they are aimed at preventing the tensions that interfere with the natural relation of body parts. And because we can't bring this condition about by releasing or relaxing muscles directly, we must focus not on the muscles themselves but on the relations of parts and wait for the muscles to release on their own.

Although the semi-supine position provides almost complete body support, there is one exception to this—namely the legs, which tend to flop outward when we let go in the thigh muscles. To counter this, we have to think of sending the knees up to the ceiling so that, when the thigh muscles release, the knees go upward instead of flopping. How this works is not immediately obvious and will take time, but as the other directions work, this will become clearer. The point is that all the directions begin to work together so that muscular release in different parts of the body becomes integrated into a whole, which in turn informs the parts.

The Principle of Non-Doing

Although many of us believe we can address tension by stretching or actively trying to relax our muscles, it is essential to recognize at the outset that we cannot establish proper muscle length by performing actions, correcting what we think is wrong, assuming postures or positions, or by forcing the body in any way. We must learn to be quiet and to stop doing anything so that the body can naturally right itself, which it is designed to do if we give the muscles a chance to lengthen. This requires an attitude of non-doing.

But what is non-doing? Sometimes people speak of non-doing as if it is a form of relaxation, which of course is true to the extent that we are stopping and giving ourselves time to do nothing. But the goal of practicing non-doing is not simply to relax or quiet down but to give muscles a chance to release into length, which is quite different from relaxation because it requires a certain kind of energy, a toning up of muscles that cannot be achieved through relaxation.

But there is another reason that non-doing is crucial, and that is that unless we stop our habitual activity, our muscles cannot release into length. We think that when we are relaxing or doing nothing, our muscles are inactive, when in fact they are often contracting unconsciously. The only way to prevent this is by coming to a full stop and, by lying or sitting in a supportive position, to remain quietly alert, to allow the muscles to let go, and to make sure we are doing nothing, so that this muscular activity can cease.

This cessation of activity, however, cannot happen immediately but takes time. When we lie down, at first it will seem as if nothing is happening because muscles that are chronically shortened do not want to let go. It is only when we have been lying quietly for some days, patiently asking for things to happen, that muscles will begin to release; when this happens, we begin to realize that we are actively tightening muscles and that our job is to stop doing this. This requires an attentive attitude because whenever we become distracted, our muscles will begin to tighten again, and our job is to notice this and to remind them to let go.

Anyone who has engaged in this practice knows that non-doing is a subtle art that runs counter to all forms of doing, bodywork, and exercise. When we work on muscles, stretch or play with them, we aren't really stopping. We may get release of some kind, we may produce changes that make us feel better, but the underlying chronic activity that keeps muscles from truly releasing, from letting go into length and from allowing the postural system to work as a reflex system, will persist. To overcome this chronic, unconscious activity, we have to make sure that we stop worrying, holding, and tightening in muscles. For many of us, this is a difficult step to take because we want to change things, to work at things, to do something to make things better, whereas non-doing requires that we stop trying to change things and, instead, allow things to work entirely by themselves. This is a practice that takes time; if we don't do it every day and with real clarity of purpose, we can't expect to command the working of the postural system as the foundation upon which this work is based.

The Semi-Supine Position

Although we are designed to be upright, the curves of our upright posture, as we've seen, are susceptible to shortening—particularly the lower back and the neck, which become chronically shortened. Because the semi-supine position provides so much support for the body, it is ideal for letting go of these tensions, for restoring lengthened muscles throughout the body, and for reestablishing the natural working of the postural system.

Some relaxation methods advocate lying flat on the floor with legs extended and no support for the head. Lying in this way, however, is problematic for two reasons. First, when the legs are fully extended, they tend to pull on the pelvis, which exaggerates the lumbar curve; at the same time, the head is thrown back, which causes the neck muscles to shorten. Second, this position tends to encourage a kind of collapse of the entire system, when in fact the body needs a kind of lively support. This is why it is important to put the knees up and to place a pile of books under your head. Putting the knees up reduces the lumbar curve of the spine, and supporting the head with books helps the neck muscles to lengthen and prevents the pulling back of the head. Both these things encourage lengthening and release of the back, leg, and neck muscles.

The Surface

It is best, when doing the semi-supine exercise, to lie on a carpet or on a wooden floor because this type of surface will provide good support for your head, your upper back, and your pelvis. Do not lie on a bed or very plush carpet because the pelvis and lower back will collapse, which will defeat the purpose of the exercise.

The Position

Place a small pile of books under your head. The books should be high enough to prevent the neck muscles from being shortened and the head from being pulled back, but not so high that the throat is pinched, the neck is stretched, or the head is held up too high. Bend your legs at the knees, with your feet fairly close to your buttocks and shoulder width apart.

The Directions

See to it that you are not tightening the muscles at the back of the neck. It helps if you think of allowing the books to support your head so that you are not pulling your head onto the books. Also, make sure you are not holding your head off the books, so that you are not tightening the neck muscles and are instead allowing the books to fully support your head (see page 59).

Think of allowing your head to come out of your back so that your back and spine can lengthen. Do not try to move your head; simply allow this to happen. As you release your neck muscles, it may feel as if your head is starting to come out of your back; at the same time, your back will tend to lengthen and to make more complete contact with the floor. It will also tend to broaden out, which means that the muscles that tend to narrow the back are releasing.

See to it that you are not tightening in your thighs and hips so that your knees can come out of your hips and back. It helps if you think of the knees pointing toward the ceiling. You may find in this position that your legs will tend to flop outward. Don't let this happen, and instead think of the knees pointing to the ceiling and think of letting go in this direction, or into length. As the other parts of the body work better, the thighs will let go as a continuation of the lengthening of the back, and the knee direction will become part of a total pattern.

It is important to remember that lying down in semi-supine is not a relaxation exercise—that is, it is not meant to bring about a complete relaxation of muscular activity but a toned, lengthened condition of the muscles. To this end, you must remain alert throughout the process and try not to space out. If you find you are getting sleepy, take a rest, but do not confuse resting with lying down in semi-supine, which is a discipline that requires an alert, aware state of mind. The semi-supine exercise is essentially a mindfulness practice, but one with a very definite physiological goal.

Getting Up from the Semi-Supine Position

When you have finished doing the semi-supine, don't get up suddenly, and don't get up by doing a sit-up, which will cause the flexors on the front of the body to violently tense up, defeating the purpose of the exercise. Instead, roll gently onto your side, taking your time and using the books to support your head like a pillow so that you will now be lying comfortably on your side. Begin to face down to the ground and get your arms beneath you so that, instead of using your abdominal muscles and tightening in your ribs to get up, you can gently push yourself up into a sitting or crawling position. Then rise slowly onto your feet.

When you are standing, take a moment to come to your full stature. Notice your breathing, your legs, your back, and your neck and shoulders. Do you feel different? Can you notice any changes in your body? Are you calmer?

The Primary Directions

When we perform actions, we tend to tighten muscles and interfere with the natural working of the muscular system. By projecting mental orders, or "directions," we prevent these tensions and bring about a more coordinated working of the body.

You can "direct" body parts in a number of ways, such as "lengthen the fingers," or "brighten the eyes." You can also work with directions in many different ways, as when we focus on a part of the body, or explore particular kinds of movements. There are, however, four main directions that relate specifically to the postural system. Because these directions are so central to the working of the musculoskeletal system, it's important to think about them and what they mean; with time, they take on more and more meaning, and begin to work in a very definite way. The directions are:

Let the neck be free.

To let the head go forward and up.

To let the back lengthen and widen.

To let the knees go forward and away.

1. Let the neck be free.

Notice that, although we pull back the head, the first direction is not the forward and up of the head, but "let the neck be free." This is because when we pull back the head, we tighten the neck to do it; we must first stop tightening the neck—particularly in the muscles at the nape of the neck that pull the head back—if we want the head to go forward and up.

2. To let the head go forward and up.

When we are sitting or standing and tighten the neck muscles, we tend to pull the head back and down. That's what the "forward and up" refers to. By releasing the muscles at the back of the neck, the pulled-back head releases to tilt forward: That's the "forward." The overall release of the muscles of the neck and trunk allows the spine to lengthen and the head to go up: That's the "up." When you're lying in the semi-supine position, the tendency will be to tighten your neck muscles and pull your head back onto the books so that it can't lengthen out of your body. So instead of thinking "forward and up," you can simply think of not pulling your head back so that it can come out of your back.

3. To let the back lengthen and widen.

Along with pulling back the head, we tend to shorten in stature by tightening and arching the back. "Lengthening" the back muscles means to allow them to regain their natural elasticity and tone; "widening" means to allow the back to broaden and for the ribs to move freely.

4. To let the knees go forward and away.

Even if we are not aware of it, when we pull the head back and shorten in stature, we also tighten in the thighs and sink into the legs. "Knees forward" means to stop tightening in the thighs so that the leg muscles lengthen and the knees come out of the back; "away" means to stop tightening in the inner thighs so that the knees come apart.

Obviously, these directions are meant to work together. That's why the four primary directions are joined together by the word "to": each one is linked to the other, to be given "one after the other, and all together." Also keep in mind that the directions are not voluntary movements, but muscular releases that happen by themselves if you allow them to—hence, the words "to let": You are letting these changes happen, not making them happen.

Kinesthetic Thinking: The Key to Directing

What does it mean to "think" the directions? When we conceive of muscle activity, or the ability to influence muscle tone, we normally think in terms of actions—like raising the arm—that produce a definite contraction, or tensing, of the muscle. Being asked to simply "think" of allowing the head to go forward and up seems, in contrast, vague and intangible. And yet it is possible to affect muscle tension just as concretely by thinking as by actively doing something. John Basmajian, one of the early biofeedback researchers, attached very fine electrodes to muscle fibers, amplified the responses, and hooked the electrodes up to an oscilloscope in order to visually display the nerve impulses on a screen. He found that when subjects could observe the oscilloscope and therefore had feedback about activity in their muscles, they could learn to consciously control them with only a few minutes of practice. Some subjects, after just a few minutes, were able to gain this control without the benefit of the oscilloscope by relying only on their own kinesthetic feedback.

This ability to consciously affect muscle activity graphically demonstrates the power of thought to concretely influence muscle tone. If the head is pulled back and the muscles at the nape of the neck are tightened, then by mentally directing the head forward and up, it is possible to release the tension in the nape of the neck and restore the natural poise of the head simply with thought. In Basmajian's biofeedback studies, subjects learned to control specific parts of specific muscles. But by directing the head in relation to the torso, this conscious power can be utilized to control not specific muscle fibers, but the overall balance of tension governed by the head and torso. If the body is collapsed, "directing" the head in this way "energizes" the muscular system by restoring length; if muscles are overly contracted, directing the head releases excess tension. In this way, conscious "direction" results in a tangible change in muscle tone throughout the body, restoring natural muscle length by removing the unnecessary tension that has interfered with the postural system. And this result is achieved, without a teacher's help, simply by "thinking."

But none of this happens without the intention, or wish, and without actually working at it. You can think of your head going "forward and up," but this isn't going to make your head go forward and up any more than thinking about the Empire State Building will make your head go forward and up. You have to want it to happen, you have to ask it to happen, and you have to see to it that it does happen. You have to actually be aware of yourself and want the directions to work; you have to actually spend time thinking. Directing is a conscious process, and you have to work at it in order for the directions to take on their full meaning.

The key to release of muscle tension, then, is the process of inhibiting muscle tension and wishing, which we do in conjunction with mechanical support. As we've seen, this is not at all the same as mechanical stretching, which simply forces connective tissue to lengthen but does not bring about internal release of muscles that are chronically tightening. When we direct, we are asking motor nerves to stop firing so that muscles can release into length. Thinking and releasing in this way is not a relaxation exercise, since we are not simply asking muscles to become flaccid or deadened but to let go into lengthened support, which increases muscle tone and activates a larger postural response.

This also cannot happen if you get sleepy or heavy. As anyone knows who has fallen asleep on a long train ride and begins to nod off, when we lose consciousness, the postural muscles that support the head and other parts of the body turn off. In the same way, if we are inattentive, heavy, or relaxed, the muscular system goes into sleep mode and stops working. For this reason, it is crucial to stay alert while practicing the semi-supine exercise, because it is only when we are fully awake and alert that the musculoskeletal system can be fully activated.

Wishing, Attending, and Non-Doing

Because the subject of muscles is so close to us, we have all sorts of strange ideas when it comes to muscles, usually based on the belief that there is something wrong in the form of deep tension, emotional holdings, or carrying stress in the body. Because of this, most us find it quite difficult to think constructively about ourselves. If, for instance, I lie down to give my directions and my back hurts, I'm likely to begin thinking about my back and to focus on what seems to be wrong with it. This, of course, is the wrong attitude, because if you direct in order to solve your back problem or in order to reduce specific tensions or because you think there is something wrong with your body and you want to correct it, you're going to get into all sorts of problems.

We have to understand, first of all, that our muscles work automatically as part of a total system, and we won't bring about an improved functioning of this total system by working with parts, but only by thinking about the whole. We've got to get away from the idea that our objective is to correct specific problems; we are working with a system designed to work on its own, and if we try to do anything to correct the system, this can't happen. We also can't help matters by trying to relax or adjust specific parts of the body because we don't really know what's wrong and what to do about it. We have to be patient and, as a starting point, stay out of our own way.

Now of course this sounds simple enough, but when it comes to this subject, even people who are otherwise quite disciplined find it difficult to stay on track and end up thinking about what they are worried about, not what they've decided to think about. When we approach the problem in this way, we are thinking negatively, and we've got to find a way to think positively and constructively, to talk nicely to ourselves, to manage our emotions. We've got to learn to do for ourselves what a teacher does: stay on track, stay focused, take our time, and stay constructive.

One way to approach directing constructively is to start the process by being aware of your surroundings. Don't direct, don't worry about your "use"; just lie there and look at the ceiling, the colors, the details. This may seem simple and basic, but in fact few of us are able to be aware of our surroundings while we pay attention to ourselves; the moment we think about ourselves, we start to feel what is happening or lose ourselves in a train of thought, and both of these things are forms of inattention, forms of worrying and holding and interfering with ourselves. In order to think constructively about ourselves, we must first and foremost be aware of what is around us.

Now it may sound rather paradoxical to say that if you want to be aware of yourself, you must be focused on what is around you, but that is how attention works. If you focus on yourself by closing your eyes, feeling your muscles, or turning inward, you're not practicing attention but inattention. If you observe animals and young children, you will see that they are instinctively aware, interested in and open to what is around them. Attention is being aware of your surroundings, and it's a very good way to start out the process of directing—especially if you spend a lot of time engaged in the kind of work that involves narrowing your attention, such as sitting at a computer or focusing on specific tasks. The first step in thinking constructively about ourselves is to be aware of and attentive to our surroundings.

When you have given yourself some time to explore this, see if you can add kinesthetic awareness to the equation without losing awareness of your surroundings. While looking at the ceiling, for instance, can you be aware of the contact of your back on the floor? Being kinesthetically aware in this way does not require that you stop being aware of your surroundings. If being aware of your back takes you away from seeing the ceiling, then stop for a moment and go back to just looking at the ceiling, and then, without becoming too worried or concerned about it, simply add your awareness of your back to the mix. Being aware of yourself and what is around you at the same time is sometimes referred to as an "expanded field of awareness," and it is the kind of awareness that this process demands.

Now obviously, just lying there and being aware of your back isn't going to make a huge difference—or rather, it might and it might not. But even if we do not get any results, we have to begin the process of attending to ourselves in this way because, even if it doesn't produce results, it is constructive and organized. We all start with the idea that we've got to feel what our muscles are doing, or work on particular problems, and we've got to be clear that that isn't our job. We've got to organize our awareness in such a way that we aren't focused on what we think is wrong but on the organization of the whole. If, for instance, you simply notice the parts of the body that are bearing weight—the back of the head, the shoulder blades, the back of the pelvis, the feet, and the arms—while still maintaining awareness of your surroundings, you'll find that you're organizing your attention in a way that actually corresponds to the postural system. Notice that you've also been thinking constructively because you haven't begun to worry or to think about what is bothering you, or what you feel should be changed; you're simply being aware without judgment. You're also being aware in a structured way that doesn't focus on specific parts of the body but on the whole. Nothing may happen, you may not feel anything happening, but you're being constructive and kind to yourself, and that, again, is what you want.

Now if you've gotten this far and feel you're ready for the next step, you can add wishing to the equation. Obviously when we lie down, we want to be aware, but we also want to ask for certain things to happen and to generate energy for things to happen. While noticing that your head is resting on the books and your back on the floor, see to it that you're not stiffening your neck or arching your back so that your head can go away from your trunk and your back can lengthen and widen. You are no longer simply being aware but actually wishing for things to happen—for the neck muscles to release, for the head to go out of the back, for the back to lengthen and widen, and for the knees to release away from the hips. You are not trying to make these things happen, but you do want changes to occur. And if you're doing this while remaining aware of your surroundings and sticking to this general framework, you're doing very well, because now you're attentive and aware of yourself and, in addition, you're directing in an organized, constructive way, and if you stick to this process, things will start to happen.

There's one more piece I'd like to add, and then we'll sum up this part of our directing discussion. If you engage in the process we've been talking about, at some point things will start to change. You may find that you were holding in your inner thighs and are now able to let them go, or you may have been arching your back or holding your ribs and can now allow the back to widen and the ribs to let go. When this happens, parts of the body will move, and when parts of the body move, this will draw your focus; you will feel the changes happening, and inevitably you will want to help. There is nothing bad about this—in fact, it is to be expected. But when it happens, you've got to see to it, first, that you don't become too distracted and instead continue to be aware of your surroundings. And you've also got to see to it that you leave things alone. In other words, the process of directing will tend to produce changes, which in turn will tempt you to "do" the directions; your job is to make sure you don't do them and get out of the way by continuing to notice your surroundings, to wish the directions, and to leave yourself alone.

To summarize, we have to:

a. Be aware of what is around us and maintain this attention.

b. Be aware of key body parts and think, or wish, for what we want.

c. Leave things alone as changes begin to happen.

d. Stick to this process over a period of time without worrying about results.

This, in a general way, is the kind of framework you want to give yourself when you direct. It's very much a process of managing your wish and your thought processes, keeping it all very clear and constructive so that you get release and not tension, while leaving room for things to happen on their own. And when you get distracted or lost in thought—and you will get distracted and lost in thought—this means that you are no longer thinking constructively. At that point you have to come back to the process without worrying about results and without becoming distracted, and by sticking to the structure you've given yourself. So we've got to maintain awareness by noticing if we lose attention to our surroundings. And when we are doing this and things begin to happen—and sooner or later things will start to happen—we have to leave things alone so that we don't interfere and allow whatever needs to happen to happen by itself.

Being clear about how to engage in this process is a necessary part of how we must manage our thinking and our emotional attachment to what we're doing so that we work in a way that corresponds not to what we think is the problem, but to how we are actually designed. If you can learn to do this and practice it diligently, you'll find that it's possible to quite literally restore the musculoskeletal system in a way that goes far beyond what you thought possible, and in a way that cannot be achieved by trying to employ any kind of method.

The Concreteness of Thinking

It is easy to think that, as a purely mental process, directing is rather vague and insubstantial when in fact it is a very concrete process with a very concrete goal. Let me demonstrate this with the simple example of directing the knees while lying in the semi-supine position (see Chapter 12). The legs are moved by muscles that attach at the pelvis and femur to the knees and below the knees, like guy wires. They have the power to contract, and, unless we are perfectly coordinated, that's exactly what these muscles will do—that is, become chronically tight and constricted. Most of this tension is unconscious, which means that we don't notice it at all; if we did, we wouldn't be doing it! Nevertheless, we are holding in these muscles and, instead of this, we want them to let go to whatever extent we are able so that the knees, instead of being pulled into the hips, will go away from the hips.

But how are we going to let go of tension that we don't even know is there? We have to direct the knees by letting them go, even if we don't feel much. We have to say, "There must be tension there, and I want to see to it that I'm letting it go by wishing for my knees to go up to the ceiling and for my thigh muscles not to be holding so that this can happen." We've got to wish this and we've got to keep wishing it.

When giving your directions, then, you've got to assume there are tensions present whether you notice them or not, and you've got to remember that it's your job to see to it that you stop tightening the neck so that the head can go away, that you stop arching or tightening the back or holding your breath or ribs so that the back muscles can widen and lengthen, and that you stop tightening the thighs so that the knees can go away. It's your job to ask for release and to see to it that you are wishing it and wanting it and, to the extent that you notice you are holding or interfering, to stop doing these things as the basis for letting the directions work. When you do this, the muscles will liven up and begin to release, and when they do, you'll become aware that you were holding, and you can use this awareness to feed your wish to keep letting go. And that process is what we call directing.

The point I'm trying to make is that while the concept of directing may seem indefinite and vague, this is mainly because the process of doing nothing and just "thinking" is unfamiliar to us. But when we accept that there are shortened muscle tissues and that they must be there if we have "use" problems, we then realize that there is nothing at all vague about directing and that it's intended to address a problem that's very real. If we think direction or chi or body awareness is going to help us, then we are just inventing something, because awareness alone doesn't mean very much. We must recognize, to begin with, that we have tensions and that we direct in order to stop making these tensions and in order to get our muscles to let go into length. From this process we discover that the body parts are fixed and held, and we learn to direct as a very concrete way of addressing a very concrete problem. If we simply take the teacher's word for it, we may not see the problem and we may not believe it's there. But we have to recognize that it is there, that the muscles are shortened and the system interfered with, and that we have to do something about this, which is one of the primary things directing is about.

Directing and Antagonistic Action

If you direct in the way we have just been talking about, improvements will inevitably happen. At first, these changes will be rather superficial, giving you a sense of calming down or of relaxing muscles that seem tight. If you engage in this process for an extended period, however, much deeper changes will take place. The neck and back muscles will let go, the leg muscles will release, you may feel the neck drop a bit and the back lengthen more. Perhaps you will discover that you were holding in your ribs and that, as they let go, your back will fill out and widen. Perhaps your hips will let go and you will find that your knees weren't really going away because your thigh muscles were holding, and now you are able to allow the knees to go forward and away. Any number of changes will occur, and it's hard to predict which ones need to happen or how. But sooner or later, muscles will release and the different parts will begin to work together as a whole until, finally, you will feel everything let go into length. This is a very distinct, dramatic, and wonderful feeling, and there is no mistaking it. It's as if the entire back musculature lets go so that the back fills out on the floor and becomes very fleshy and elastic, and the neck releases so that the head seems to come right out of the back and the knees go away and the whole system becomes toned and lets go into length and width.

This new change isn't merely bringing about improvements; it is establishing the elastic, antagonistic working of the muscles that we've been speaking about throughout this book. I particularly wanted to mention this because many people who practice methods for lengthening the spine or relaxing muscles will get a feeling of upward release of the head and a lightening up and lengthening of the system; they think that they're experiencing what it means to establish a coordinated working of the system when in fact it's only a hint of what it means to be coordinated. The purpose of directing, ultimately, isn't simply to bring about release here or there or a general lightening up of the system, but to activate the antagonistic action of the muscles. We can see this principle at work in young children, who are not releasing muscles or utilizing the postural system to become lighter and freer or more effortless, but display a naturally lengthened, coordinated, and elastic quality in the musculature such that the entire system is highly integrated, with the head and back clearly and elastically continuous with one another, the back full and widening, the ribs completely free, and the legs undoing and lengthening out of the back. This is a truly coordinated condition of the system, and even those of us who think we understand what this means are often rather far removed from experiencing it because we're doing too many things to interfere and have gotten ourselves too tense and collapsed and stuck to be able to regain it.

It is easy to think that because we're not trying to make anything happen and instead must focus on the process, we're not supposed to be specific about what we want, but that's just a lot of nonsense. There is something definite that we're after, and that is the antagonistic action of the muscles. When the muscles work antagonistically and are letting go so that the parts to which they attach are moving away from each other, this activates the entire reflex response of the postural system. For this to happen, various areas of the body that have become shortened and fixed have to regain their elasticity and flexibility, and various holdings have to give way to allow real stretch and undoing in muscles. At the same time, there has to be a toning up of various muscles that are not doing enough work so that the different parts of the body are connecting up into one coordinated whole. This, in fact, is the real purpose of directing, and until it's happening, it's impossible to reap the full benefit and meaning of this work that, once again, is not just about release of the neck and upward movement of the head and lengthening of the back, but a coordinated working of the entire system based on the principle of the antagonistic working of the musculature.

Let's return, for a moment, to the directing we did a moment ago. If you lie down and give your directions and see to it that you aren't tightening and shortening, things will improve, but the neck isn't going to be the first thing that will release—in fact, it's often the last. After lots of other changes happen—particularly the knees going away with release across the hips and the back lengthening and the entire trunk undoing and untwisting—when all those things happen, the neck is going to start unwinding and you're going to start feeling the head come out of the back and the knees go the other way and you're going to see how the whole thing was being interfered with, and you'll say, "My god, so that's how it works."

The point I'm trying to make is that when we direct, there is something very definite we are after. We're trying to establish a condition in which the muscles that are contracted and pulling on skeletal parts start to let go so that the skeletal parts seem to be dynamically exerting stretch on the muscles. When this happens, the muscles will fully let go: The head will release out of a lengthening back; the back will fill out and widen; instead of pulling on the skull, the throat will release and allow the skull to go up; the shoulders will spread apart across the chest; and the hips and legs will let go so that the knees will actively go away from the hips. In short, instead of muscles pulling on bones, muscles let go of the bones and all the parts let go into expansion—the true tensegrity paradox of non-effort, of doing more with less, of parts working oppositionally, of muscles working powerfully and energetically but with a minimum of effort. And because this elastic condition of the musculature reestablishes the forward-and-up balance of the head in relation to the trunk, it also triggers the automatic reflex response of the postural system so that everything begins to work by itself.

To achieve this, we have to think about and be clear on what we are trying to accomplish, and we have to work at this with our awareness, our attention, and our wish. We have to think for periods of time, maintaining both focus and a broad alertness. We must ask for a number of specific things to happen yet not focus on any one of them, and we must exert tremendous patience in getting out of the way and allowing things to happen on their own. If we do this, we will bring about a level of change we never thought possible and elicit a natural, automatic response that will give us an entirely new view of how the body works, its wondrous design, and its naturally vital state.

NOTES

Introduction

1. Alexander, *The Use of the Self*, 8–14.
2. Dimon, *Neurodynamics*, 3–49.

Chapter 1

1. Fuller, "Tensegrity," 112–127, 144, 148. An early version of a tensegrity structure was created by K. D. Snelson, who produced artworks based on the concept; see Snelson, K. D. "Continuous Tension."
2. The term *biotensegrity* was coined by Stephen Levin; see Levin, "Tensegrity-Truss as a Model," 375–388.
3. Van der Wal, "Architecture of Connective Tissue," 9–23.
4. Turvey and Fonseca, "The Medium of Haptic Perception," 145.
5. Ingber, "Tensegrity 1," 1157–1173. See also Turvey and Fonseca, "The Medium of Haptic Perception," 148–150.
6. Turvey and Fonseca, "The Medium of Haptic Perception," 149–150.
7. Roberts, "Contribution of Elastic Tissues," 266–275.
8. Colombini et al., "Non-Crossbridge Stiffness," 153–160. See also Nishikawa, "Eccentric Contraction," 189–196.
9. *New York Times*, "Slight Change in Gait," May 30, 1995.
10. Macintosh et al., "The Biomechanics of the Thoracolumbar Fascia," 80–81.

Chapter 2

1. Dimon, *Anatomy of the Moving Body*.

Chapter 3

1. The importance of the forward balance of the head on the spine and the lengthening of the trunk were first described by F. Matthias Alexander in *The Use of the Self*, published in 1932. Alexander observed that, when he vocalized, he pulled back his head and shortened in stature. To use his voice well, he discovered that he must lengthen in stature and his head must go forward and up—a relationship that he called "the primary control of my use in all my activities."

2. Kandel et al., *Principles of Neural Science*, 564–579.

3. Kandel, Schwartz, and Jessell, *Essentials of Neural Science and Behavior*, 505.

Chapter 5

1. Woodham et al., "Long-Term Lumbar Multifidus Atrophy," 27–34.

2. Gray, *Gray's Anatomy*, 349.

Chapter 6

1. See Dart, "Voluntary Musculature in the Human Body: The Double Spiral Arrangement," 68.

2. This exercise was introduced to me at the Constructive Teaching Centre in London, England, by Dilys Carrington, who used the procedure in working with students.

Chapter 8

1. In a paper titled "The Postural Aspect of Malocclusion" (pp. 1–21), Dart links the problem of malocclusion of the jaw to postural imbalances that center on the deep flexors of the neck and throat, or what he calls the *rectus cervicalis sheet*.

2. In his extensive study of the comparative anatomy of the larynx, *The Mechanism of the Larynx*, Negus demonstrates that the larynx originated as a sphincter muscle to protect the airway, and that this remains the primary function of the larynx in reptiles and mammals.

3. Husler and Rodd-Marling argue in *Singing: The Physical Nature of the Vocal Organ* that a fully functional voice requires that the larynx be actively supported within a network of muscles, which they call the *suspensory mechanism* of the larynx (pp. 24–30).

4. In a biographical account in *The Use of the Self*, F. Matthias Alexander observed that when he vocalized, he "depressed the larynx"; he connected this with a larger pattern of pulling back his head and shortening in stature (p. 13).

5. In an article titled "Glosso-Postural Syndrome," Scoppa argues that the tongue and its related apparatus can influence total posture.

BIBLIOGRAPHY

Alexander, F. Matthias. *The Use of the Self.* London: E. P. Dutton, 1942.

Colombini, Barbara, Marta Nocella, and Maria Angela Bagni. (2016). "Non-Crossbridge Stiffness in Active Muscle Fibres." *Journal of Experimental Biology* 219 (2016): 153–160.

Dart, Raymond. "The Postural Aspect of Malocclusion." *Journal of Dentistry of South Africa* (1946): 1–21.

Dart, Raymond. "Voluntary Musculature in the Human Body: The Double Spiral Arrangement." *British Journal of Physical Medicine* 68 (1950).

Dimon, Theodore. *Anatomy of the Moving Body: A Basic Course in Bones, Muscles, and Joints.* Berkeley, CA: North Atlantic Books, 2001.

Dimon, Theodore. *Neurodynamics: The Art of Mindfulness in Action.* Berkeley, CA: North Atlantic Books, 2015.

Fuller, R. Buckminster. *Synergetics: Exploration in the Geometry of Thinking.* New York, NY: Macmillan, 1975.

Fuller, R. Buckminster. "Tensegrity." *Portfolio Artnews Annual* 4 (1961): 112–127.

Gray, Henry. *Gray's Anatomy.* New York: Bounty Books, 1977.

Husler, Frederick, and Yvonne Rodd-Marling. *Singing: The Physical Nature of the Vocal Organ.* London: Hutchinson, 1976.

Ingber, Donald. E. "Tensegrity 1. Cell Structure and Hierarchical Systems Biology." *Journal of Cell Science* 116 (2003): 1157–1173.

Kandel, Eric R., James H. Schwartz, and Thomas M. Jessell. *Essentials of Neural Science and Behavior.* Stamford, Connecticut: Appelton and Lange, 1995.

Kandel, Eric R., James H. Schwartz, Thomas Jessell, and A. J. Hudspeth, eds. *Principles of Neural Science.* New York: McGraw-Hill, 2010.

Levin, Stephen M. "The Tensegrity-Truss as a Model for Spine Mechanics: Biotensegrity." *Journal of Mechanics in Medicine and Biology* 2, no. 3 (2002): 375–388.

Macintosh, J. E., N. S. Bogduk, and S. Gracovetsky. "The Biomechanics of the Thoracolumbar Fascia." *Clinical Biomechanics* 2, no. 2 (1987): 78–83.

Negus, Victor E. *The Mechanism of the Larynx.* London: Heinemann Medical Books, 1929.

New York Times. "Slight Change in Gait Makes Burden Lighter." May 30, 1995.

Nishikawa, Kiisa. "Eccentric Contraction: Unraveling Mechanisms of Force Enhancement and Energy Conservation." *Journal of Experimental Biology* 219 (2016): 189–196.

Roberts, Thomas J. "Contribution of Elastic Tissues to the Mechanics and Energetics of Muscle Function During Movement. *Journal of Experimental Biology* 219 (2016): 266–275.

Schleip, Robert, Thomas W. Findley, Leon Chaitow, and Peter A. Huijing. *Fascia: The Tensional Network of the Human Body.* Toronto, Ontario: Churchill Livingstone Elsevier, 2012.

Scoppa, F. "Glosso-Postural Syndrome." *Annali di Stomatologia* 54, no. 1 (January–March 2005): 27–34.

Snelson, K. D. Continuous Tension, Discontinuous Compression Structures. US Patent 3,169,611, Filed March 14, 1960, issued February 16, 1965.

Turvey, Michael T., and Sergio T. Fonseca. "The Medium of Haptic Perception: A Tensegrity Hypothesis." *Journal of Motor Behavior* 46, no. 3 (2014): 143–187.

Van der Wal, Jaap. "The Architecture of the Connective Tissue in the Musculoskeletal System—An Often Overlooked Functional Parameter as to Proprioception in the Locomotor Apparatus." *International Journal of Therapeutic Massage and Bodywork* 2, no. 4 (December 7, 2009): 9–23.

Woodham, Mark, Andrew Woodham, Joseph G. Skeate, and Michael Freeman. "Long-Term Lumbar Multifidus Muscle Atrophy Changes Documented with Magnetic Resonance Imaging: A Case Series." *Journal of Radiology Case Reports* 8, no. 5 (May 2014): 27–34.

Young, J. Z. *The Life of Mammals.* Oxford: Clarendon Press, 1970.

INDEX

PHOTO CREDITS

ABOUT THE AUTHOR

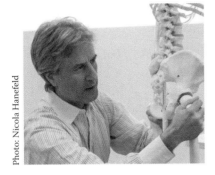

THEODORE DIMON, EdD, is an educator, philosopher, psychologist, and author who has dedicated his life to the study of mind and body and our unique human capacity for conscious development. He is a leading theorist in the field of psychophysical education, which focuses on the study of mind and body in action as an educational discipline. He is the founder and director of The Dimon Institute in New York City, an organization dedicated to training educators and to researching and studying the human being and its evolving conscious systems. Dr. Dimon has also lectured as an adjunct professor of clinical psychology and education at Columbia Teachers College.

Dr. Dimon's areas of expertise include the psychology of attention, the study of the body in action, the mastery of skill, anatomical function and design, the study of stress, the vocal and respiratory systems, and comparative evolution. He is a leading expert and scholar on the work of F. Matthias Alexander, a pioneer in the study of habit, musculoskeletal function, and the control of action.

Anatomy in Action was written by Dr. Dimon as an innovative and cutting-edge textbook for his own students at The Dimon Institute to explain how key musculoskeletal systems of the human body are designed to work both individually and holistically during movement, and how these systems can potentially fail, leading to stress and pain.

For further information, visit www.theodoredimon.com and www.dimon institute.org.

About the Illustrator

G. DAVID BROWN is a professor emeritus after fifteen years of teaching illustration at Winthrop University in Rock Hill, South Carolina, where he developed the BFA illustration program in the department of design. He has been a medical illustrator since 1977, having earned his master's degree in medical illustration from the University of Texas Health Science Center at Dallas. His undergraduate degree was in visual and environmental studies from Harvard University, concentrating in film animation and the psychology of perception. He has collaborated on a number of books with Dr. Dimon, beginning in 2011 with *The Body in Motion: Its Evolution and Design,* which won the Association of Medical Illustrators' Best Book of the Year 2011 in the category of Books for Allied Health.

ALSO BY THEODORE DIMON

available from North Atlantic Books

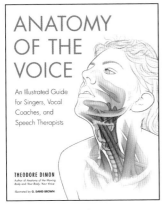

Anatomy of the Voice
978-1-62317-197-1

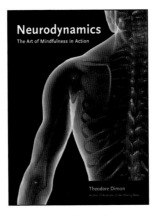

Neurodynamics
978-1-58394-979-5

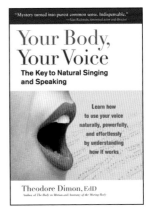

Your Body, Your Voice
978-1-58394-320-5

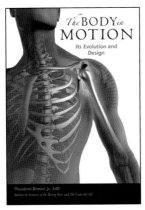

The Body in Motion
978-1-55643-970-4

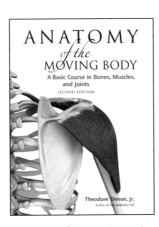

*Anatomy of the Moving Body,
Second Edition*
978-1-55643-720-5

The Elements of Skill
978-1-55643-476-1

North Atlantic Books
www.northatlanticbooks.com

North Atlantic Books is an independent, nonprofit publisher committed to a bold exploration of the relationships between mind, body, spirit, and nature.

About North Atlantic Books

North Atlantic Books (NAB) is a 501(c)(3) nonprofit publisher committed to a bold exploration of the relationships between mind, body, spirit, culture, and nature. Founded in 1974, NAB aims to nurture a holistic view of the arts, sciences, humanities, and healing. To make a donation or to learn more about our books, authors, events, and newsletter, please visit www.northatlanticbooks.com.